T0076212

Molecular Genetics of Dysregulated pH Homeostasis

Jen-Tsan Ashley Chi
Editor

Molecular Genetics of Dysregulated pH Homeostasis

Editor
Jen-Tsan Ashley Chi
Department of Molecular Genetics
and Microbiology
Duke Center for Genomic and Computational
Biology
Durham
North Carolina
USA

ISBN 978-1-4939-1682-5 ISBN 978-1-4939-1683-2 (eBook)
DOI 10.1007/978-1-4939-1683-2
Springer New York Heidelberg Dordrecht London

Library of Congress Control Number: 2014951518

Printed on acid-free paper

Springer is part of Springer Science+Business Media (www.springer.com)

Contents

1 Introduction: Molecular Genetics of Acid Sensing and Response 1
Chao-Chieh Lin, Melissa M. Keenan and Jen-Tsan Ashley Chi

Part I "Sensing Acidity"

2 The Molecular Mechanism of Cellular Sensing of Acidity 11
Zaven O'Bryant and Zhigang Xiong

3 The Molecular Basis of Sour Sensing in Mammals 27
Jianghai Ho, Hiroaki Matsunami and Yoshiro Ishimaru

4 Function and Signaling of the pH-Sensing G Protein-Coupled Receptors in Physiology and Diseases 45
Lixue Dong, Zhigang Li and Li V. Yang

Part II "Response to Acidity"

5 Response to Acidity: The MondoA–TXNIP Checkpoint Couples the Acidic Tumor Microenvironment to Cell Metabolism 69
Zhizhou Ye and Donald E. Ayer

6 Regulation of Renal Glutamine Metabolism During Metabolic Acidosis 101
Norman P. Curthoys

7 Extracellular Acidosis and Cancer 123
Maike D. Glitsch

8 A Genomic Analysis of Cellular Responses and Adaptions to Extracellular Acidosis 135
Melissa M. Keenan, Chao-Chieh Lin and Jen-Tsan Ashley Chi

Index 159

Contributors

Donald E. Ayer Huntsman Cancer Institute, Department of Oncological Sciences, University of Utah, Salt Lake City, UT, USA

Jen-Tsan Ashley Chi Department of Molecular Genetics and Microbiology, Center for Genomic and Computational Biology, Duke Medical Center, Durham, NC, USA

Norman P. Curthoys Department of Biochemistry and Molecular Biology, Colorado State University, Fort Collins, CO, USA

Lixue Dong Department of Oncology, Brody School of Medicine, East Carolina University, Greenville, NC, USA

Maike D. Glitsch Department of Physiology, Anatomy and Genetics, University of Oxford, Oxford, England

Jianghai Ho Department of Molecular Genetics and Microbiology, Duke University Medical Center, Durham, NC, USA

Department of Neurobiology, Duke University Medical Center, Durham, NC, USA

Yoshiro Ishimaru Department of Applied Biological Chemistry, Graduate School of Agricultural and Life Sciences, The University of Tokyo, Bunkyo-ku, Tokyo, Japan

Yoshiro Ishimaru Department of Applied Biological Chemistry, Graduate School of Agricultural and Life Sciences, The University of Tokyo, Bunkyo-ku, Tokyo, Japan

Melissa M. Keenan Department of Molecular Genetics and Microbiology, Center for Genomic and Computational Biology, Duke Medical Center, Durham, NC, USA

Zhigang Li Department of Oncology, Brody School of Medicine, East Carolina University, Greenville, NC, USA

Chao-Chieh Lin Department of Molecular Genetics and Microbiology, Center for Genomic and Computational Biology, Duke Medical Center, Durham, NC, USA

Hiroaki Matsunami Department of Molecular Genetics and Microbiology, Duke University Medical Center, Durham, NC, USA

Department of Neurobiology, Duke University Medical Center, Durham, NC, USA

Zaven O'Bryant Department of Neurobiology, Morehouse School of Medicine, Atlanta, GA, USA

Zhigang Xiong Department of Neurobiology, Morehouse School of Medicine, Atlanta, GA, USA

Li V. Yang Department of Oncology, Department of Internal Medicine, Department of Anatomy and Cell Biology, Brody School of Medicine, East Carolina University, Greenville, NC, USA

Zhizhou Ye Huntsman Cancer Institute, Department of Oncological Sciences, University of Utah, Salt Lake City, UT, USA

Chapter 1
Introduction: Molecular Genetics of Acid Sensing and Response

Chao-Chieh Lin, Melissa M. Keenan and Jen-Tsan Ashley Chi

Introduction

Most biological reactions and functions occur in body fluid within narrow ranges of proton level around neutral environments. Slight changes in the pH environment have great impacts on the biological function at every level, including protein folding, enzymatic activities, cell proliferation, and cell death. Therefore, maintaining the pH homeostasis at the local or systemic level is one of the highest priorities for all multicellular organisms. When the pH homeostasis is disrupted in various physiological adaptations and pathological situations, the resulting acidity alters the cellular physiology, metabolism, and gene expression as active participants in the pathophysiological events and modulates disease outcomes. Therefore, understanding how various cells sense and react to pH imbalance through the "acid sensor" have broad impact in a wide variety of human diseases, including cancer, stroke, myocardial infarction, diabetes, and renal and infectious diseases.

Over the years, many attempts have been made to identify the acid sensor and "acid-induced factors" in different cell types, but no master acid sensor and response have been identified so far. Instead, at least three levels of complexity in the acid sensing and response is becoming clear. First, a wide variety of proteins respond to the acidity through specific acid-sensing receptors or nonspecific pH-sensitive alterations. Each of the protein or a group of proteins results in distinct downstream events and biological pathways to comprise the complex signaling and biological acidosis response. Second, different concentrations of protons and degrees of acidity may trigger different acid-sensing receptors and mechanisms to mediate distinct quantitative and qualitative acidosis response. Third, different cell types are exposed to varying pH ranges and have different sets of protein expression. Therefore, the acid-sensing mechanisms and responses to different proton concentrations are likely to vary significantly among different cell types. In this book, we have invited many experts to highlight various aspects of the molecular genetics on how

J-T. A. Chi (✉) · C.-C. Lin · M. M. Keenan
Department of Molecular Genetics and Microbiology, Center For Genomic and Computational Biology, Duke Medical Center, Durham, NC 27708, USA
e-mail: jentsan.chi@duke.edu

J-T. A. Chi (ed.), *Molecular Genetics of Dysregulated pH Homeostasis*,
DOI 10.1007/978-1-4939-1683-2_1

mammalian cells sense and respond to acidosis and their implications in the normal physiological adaptations and pathogenesis.

Acidity as Environmental Cues and Stimuli

High proton levels and acidity convey important cues for environmental stimuli. For example, the sour taste is stimulated by acidity and an increase of the proton concentration on the surface of the tongue. Excessive protons and acidity interact with the chemosensory apical membrane of taste cells to trigger the sensing of "sourness" of the food. The identification of the polycystic kidney disease 2-like 1 protein (PKD2L1) as sour receptor was first reported using the reconstitution systems to identify the ligands for G protein-coupled receptor (GPCR). The identification of sour-selective taste cells was introduced with the finding that those taste cells that express the protein PKD2L1 are necessary for sour taste in mice. Genetically driven ablation of PKD2L1-expressing cells specifically removed the sour taste, whereas the other taste qualities persisted. Here, Dr. Ishimaru and Dr. Matsunami have provided an excellent review of the molecular mechanisms of sour taste sensing to illustrate how acidity may provide environmental cues and properties of the food.

The Pathogenesis of Acidosis and Lactic Acidosis

Acute blockage of blood vessels or chronic imbalance between blood perfusion and oxygen consumption in human body can lead to hypoperfusion (lack of adequate blood perfusion) and tissue hypoxia the resulting dysregulation of pH homeostasis. The dysregulated pH homeostasis is often exhibited as excessive proton (acidosis), especially in the form of lactic acidosis. The lactic acidosis is caused by the anaerobic metabolism of glucose, which promotes animal cells to produce lactate, adenosine triphosphate (ATP), and water. The free proton is generated when ATP is hydrolyzed to adenosine diphosphate (ADP) and inorganic phosphate (Pi) and released to cause acidosis. Both ADP and Pi are also efficient substrates for anaerobic glycolysis. Every mole of glucose, when metabolized anaerobically, produces to 2 mol of lactate and 2 mol of protons, which were buffered by various buffer systems in the cells and human body. When oxygen is available for oxidative phosphorylation, extra protons can enter the mitochondria and are used for oxidative phosphorylation. Whenever production of lactate and proton exceeds the utilization and buffer capacity, it can result in lactic acidosis.

In response to acidosis or lactic acidosis, several homeostatic mechanisms are triggered at the cellular and organismic levels to limit further lactate production and enhance utilization as compensatory mechanisms to alleviate acidosis. First, intracellular acidosis inhibits 6-phosphofructokinase, one of the key enzymes in glucose metabolism, to reduce glycolysis and production of lactic acidosis. Second, lactic acidosis activates MondoA-Mlx to trigger the expression of thioredoxin-interacting protein (TXNIP) that blocks the glucose uptake by phosphorylation of glucose transporter 1 (GLUT1). Third, lactic acidosis also inhibits the oncogenic pathways of protein kinase B/phosphatidylinositide 3-kinase (Akt/PI3K). Some of these regulations are well discussed in the chapter of Dr. Ayer in the context of MondoA–TXNIP as a novel metabolic checkpoint under stresses. Moreover, the kidney also plays an important role of disposing lactate and excessive protons. Acidosis increases the activities and mRNA stability of glutaminase (GA) and phosphoenolpyruvate carboxykinase (PEPCK) mRNAs. Increased renal catabolism of plasma glutamine during acidosis generates two ammonium ions that facilitate the excretion of acids. The pH-responsive increase in PEPCK enhances gluconeogenesis and helps to remove lactate by the kidney. This metabolic adaptation of renal epithelial cells to acidosis is nicely summarized by Dr. Curthoys in the chapter on how the acidosis affects the glutamine metabolisms.

The Molecular Mechanisms of Sensing Acidosis

Given the importance of acidosis, various cells have developed sophisticated mechanisms to sense the extracellular acidosis. First, extracellular acidosis may alter the extracellular and intracellular biochemical milieu by affecting the protonation status of amino acids and proteins to alter the functional status of many cellular proteins. Among all the amino acids, histidine is the only H^+ titratable residue within the physiological pH ranges that occurs during physiological and pathological conditions. Therefore, the histidine residues of many proteins can alter their conformations and are implicated as pH sensors in many proteins. For example, acidosis inhibits the enzymatic activities of 6-phosphofructokinase and lactate dehydrogenase to reduce glycolysis, resulting in production of lactate.

Second, evidences are accumulating for the important role of membrane acid-sensing receptors in the cellular acidosis responses. These acid-sensing receptors mostly belong to two protein families: GPR4 family of GPCRs and acid-sensing ion channels (ASICs). These two families of proteins respond to very distinct pH ranges: while acid-sensing GPCRs have a pH 50 % of 6.5, ASICs have a pH 50 % of around 5.5. ASICs are proton-gated, amiloride-sensitive, voltage-insensitive cation channels belonging to the degenerin/epithelial sodium channel (DEG/ENaC) superfamily of ion channels. Given the measured intratumor pH is around 6.5–6.9 and the pH of 6.7 in our acidosis response, GPCR may be more relevant for the acidosis response in tumors. In the acute ischemia conditions of stroke and ischemic cardiac diseases, the tissue pH can drop down to 5–5.5. Therefore, ASICs are likely to play an important role in the cellular damages and death under these acute ischemic events. In this book, Dr. Zhigang Xiong has contributed a chapter to summarize the role of ASICs and other acid-sensing mechanisms in the ischemic diseases.

The acid-sensing GPCR family includes four closely related members: (1) G2A (G2 accumulation) [1], (2) GPR68 (OGR1, ovarian cancer GPCR) [2], (3) GPR65 (TDAG8, T cell death-associated gene 8) [3], and GPR4 [4, 5]. These proteins are multifunctional receptors which respond both to extracellular acidosis (proton sensing) and various lysolipid molecules as their natural agonist and antagonist ligands [4]. The activation of these acid-sensing GPCR stimulates an increase in the intracellular cyclic adenosine monophosphate (cAMP) and inositol triphosphate (IP3) with different degree of sensitivity to various concentrations of protons [6]. The basis of the acid sensing is due to the destabilization of the hydrogen bonds in several histine residues in these GPCRs under acidosis and leads to active conformation to trigger the downstream signaling. Importantly, these endogenous ligand lysolipids can directly interact with the proton-sensing functions of these receptors and can serve as either an antagonist or agonist for the acid-sensing receptors [5, 7]. These observations suggest that lysolipids and their synthetic variants may be used to modulate the cellular acidosis responses.

Although initially thought to have rather limited tissue expression, these acid-sensing receptors were later found to be expressed in many tissues, including benign and cancerous epithelial tissues [5, 8]. This wide tissue distribution of these acid-sensing receptors strongly suggests their probable roles in the acidosis response in many cell types, including different epithelial cancerous and endothelial cells. In this book, Dr. Yang has contributed a chapter to summarize the role of acid-sensing GPCR, especially in the context of cardiovascular systems.

Whether acidosis can promote oncogenesis is still debating. While certain studies suggest that acidosis response can inhibit oncogenic transformation [9], we are also aware of other studies which suggest that the acidosis response or acid-sensing receptor may be oncogenic [10] or select for aggressive phenotypes [11–13]. This contradiction and complexity may reflect the cell-type specific responses to acidosis, the involvement of different acid-sensing receptors as well as short-term versus long-term consequences of acidosis. For example, sphingosylphosphorylcholine (SPC), a natural glycolipid ligand, has entirely opposite effects on different proton-sensing GPCR—while activation of OGR1 lead to the inhibition of cell growth [14, 15], similar activation of GPR4 stimulate cell growth and migration [16] instead. Therefore, understanding the spatio-temporal patterns of extracellular protons and how these distinct downstream signals from different acid-sensing receptors integrate to generate the overall response of a given cell may provide insight to identify therapeutic targets and provide predictions to improve treatment strategies, especially when these acid-sensing receptors belong to GPCR, a protein family known to be very "druggable" and likely to be modulated by small molecule compounds. In this book, Dr. Glitsch has contributed a chapter to explain how extracellular proton concentration can affect cells in cancerous tissue by interacting with different acid-sensing receptors.

Integrative Genomic Approaches to Identify the Somatic Mutations Selected by Lactic Acidosis and Hypoxia in Human Cancers

Even though lactic acidosis is a prominent feature of solid tumors, we have limited understanding about how lactic acidosis influences the genetic, epigenetic, proteomic, and metabolic phenotypes of cancer cells. In addition to many single-gene and hypothesis-driven studies, various "-omics" approaches have been used to define the transcriptional, metabolomic, and proteomic responses of cancer cells' responses to acidosis or lactic acidosis. We have summarized the current studies of these approaches to lactic acidosis and how these in vitro studies are related to the in vivo tumor phenotypes.

One important aspect of oncogenesis is genomic instability and high frequency of somatic mutations. Cells bearing certain mutations obtain survival or proliferation advantages over cells not having these mutations. Over the extended time of tumor developments, cancer cells with these mutations expand clonally to become the dominant component of tumors, a process termed somatic evolution [17–21]. The chronic presence of tumor microenvironmental stresses, including lactic acidosis, has been proposed to serve as an important factor in the selection of cancer cells during somatic evolution. This concept is illustrated by studies showing that hypoxia enriches for tumor cells that lack p53 [22] and glucose deprivation selects for tumor cells that bear Kirsten rat sarcoma (KRAS) mutations [23]. Such selection of somatic mutation by each stress condition may be related to their distinct effects on cancer cells and resulting adaptive responses. For example, cells with KRAS mutation may be selected for under glucose deprivation because the mutation is allowing for increased glucose uptake [23, 24]. In addition, these stresses are known to induce further genomic instability and increase gene amplification [25–27]. To capture and integrate the influences of these stresses, many elegant mathematic models have been described to account for their effects on tumor progression [17, 28–31].

While driver mutations in circulating nucleic acids (CNAs) are often assumed to confer growth advantages due to proliferation or reduced cell death, we reason that some genes in CNAs are essential for survival only under stresses in a synthetic lethality relationship and offer survival advantages under tumor microenvironmental stresses. Synthetic lethality arises when a combination of mutations in two or more genes leads to cell death, whereas a mutation in only one of these genes does not. In the context of tumor microenvironmental stresses, we use the term synthetic lethality to describe genes that are essential for survival under stressed but not control conditions. This concept can be illustrated by our findings in preliminary data that activating transcription factor 4 (*ATF4)* amplification in a subset of breast cancer cells lead to high levels of an *ATF4*-driven gene expression program and provide a survival advantage under combined hypoxia and lactic acidosis [32]. These copy number alterations at the DNA levels often lead to the coordinated over- or under-expression of genes in the amplified and deleted regions. Therefore, can be used to the gene expression data of stress-response genes to identify these CNAs that may reflect the selection processed under tumor microenvironmental stresses [33]. The use of large-scale RNA interference (RNAi) or overexpression functional genomics screens in cells experiencing these stresses will allow systematic identification of genes and mutations critical for cell survival under these stresses.

Although the concept of selective pressure by these stresses is well recognized, there is a gap between the biological understanding of somatic evolution under stresses and the actual somatic mutations observed in various human cancers. While many current studies on somatic evolution under stresses have focused on the roles of glucose utilization or angiogenesis, genes in other biological processes could also be involved in the somatic evolution under stresses. These genes can be only efficiently identified through an unbiased genomic screening approach. Furthermore, integrative analysis of the global functional genomic screen with the CNAs associated with stress phenotypes has the potential to identify those somatic mutations that may have been specifically selected for by these stress conditions. In this book, Dr. Chi has summarized that how to bridge the gap between the concepts of somatic evolutions of tumor cells under stresses and the resulting CNAs seen in human cancers. Understanding of the survival mechanisms under stresses and how CNAs may provide hardwired advantages and circumvent the barriers of stresses may may enable scientists to target such survival mechanisms for therapeutic potential.

Acknowledgments This project is supported by NIH CA125618, CA106520, and the Department of Defense W81XWH-12–1-0148 and W81XWH-14-1-0309.

References

1. Murakami N, Yokomizo T, Okuno T, Shimizu T (2004) G2A is a proton-sensing G-protein-coupled receptor antagonized by lysophosphatidylcholine. J Biol Chem 279(41):42484–42491
2. Ludwig MG, Vanek M, Guerini D, Gasser JA, Jones CE, Junker U, Hofstetter H, Wolf RM, Seuwen K (2003) Proton-sensing G-protein-coupled receptors. Nature 425(6953):93–98
3. Wang JQ, Kon J, Mogi C, Tobo M, Damirin A, Sato K, Komachi M, Malchinkhuu E, Murata N, Kimura T et al (2004) TDAG8 is a proton-sensing and psychosine-sensitive G-protein-coupled receptor. J Biol Chem 279(44):45626–45633
4. Tomura H, Mogi C, Sato K, Okajima F (2005) Proton-sensing and lysolipid-sensitive G-protein-coupled receptors: a novel type of multi-functional receptors. Cell Signal 17(12):1466–1476
5. Im DS (2005) Two ligands for a GPCR, proton vs lysolipid. Acta Pharmacol Sin 26(12):1435–1441
6. Radu CG, Nijagal A, McLaughlin J, Wang L, Witte ON (2005) Differential proton sensitivity of related G protein-coupled receptors T cell death-associated gene 8 and G2A expressed in immune cells. Proc Natl Acad Sci U S A 102(5):1632–1637
7. Mogi C, Tomura H, Tobo M, Wang JQ, Damirin A, Kon J, Komachi M, Hashimoto K, Sato K, Okajima F (2005) Sphingosylphosphorylcholine antagonizes proton-sensing ovarian cancer G-protein-coupled receptor 1 (OGR1)-mediated inositol phosphate production and cAMP accumulation. J Pharmacol Sci 99(2):160–167
8. Sin WC, Zhang Y, Zhong W, Adhikarakunnathu S, Powers S, Hoey T, An S, Yang J (2004) G protein-coupled receptors GPR4 and TDAG8 are oncogenic and overexpressed in human cancers. Oncogene 23(37):6299–6303
9. Weng Z, Fluckiger AC, Nisitani S, Wahl MI, Le LQ, Hunter CA, Fernal AA, Le Beau MM, Witte ON (1998) A DNA damage and stress inducible G protein-coupled receptor blocks cells in G2/M. Proc Natl Acad Sci U S A 95(21):12334–12339
10. Zohn IE, Klinger M, Karp X, Kirk H, Symons M, Chrzanowska-Wodnicka M, Der CJ, Kay RJ (2000) G2A is an oncogenic G protein-coupled receptor. Oncogene 19(34):3866–3877
11. Gatenby RA, Gillies RJ (2004) Why do cancers have high aerobic glycolysis? Nat Rev Cancer 4(11):891–899
12. Laconi E (2007) The evolving concept of tumor microenvironments. Bioessays 29(8):738–744

13. Shi Q, Le X, Wang B, Abbruzzese JL, Xiong Q, He Y, Xie K (2001) Regulation of vascular endothelial growth factor expression by acidosis in human cancer cells. Oncogene 20(28):3751–3756
14. Xu Y, Zhu K, Hong G, Wu W, Baudhuin LM, Xiao Y, Damron DS (2000) Sphingosylphosphorylcholine is a ligand for ovarian cancer G-protein-coupled receptor 1. Nat Cell Biol 2(5):261–267
15. Xu Y, Fang XJ, Casey G, Mills GB (1995) Lysophospholipids activate ovarian and breast cancer cells. Biochem J 309(Pt 3):933–940
16. Zhu K, Baudhuin LM, Hong G, Williams FS, Cristina KL, Kabarowski JH, Witte ON, Xu Y (2001) Sphingosylphosphorylcholine and lysophosphatidylcholine are ligands for the G protein-coupled receptor GPR4. J Biol Chem 276(44):41325–41335
17. Gatenby RA, Smallbone K, Maini PK, Rose F, Averill J, Nagle RB, Worrall L, Gillies RJ (2007) Cellular adaptations to hypoxia and acidosis during somatic evolution of breast cancer. Br J Cancer 97(5):646–653
18. Gao C, Furge K, Koeman J, Dykema K, Su Y, Cutler ML, Werts A, Haak P, Vande Woude GF (2007) Chromosome instability, chromosome transcriptome, and clonal evolution of tumor cell populations. Proc Natl Acad Sci U S A 104(21):8995–9000
19. Nowell PC (1976) The clonal evolution of tumor cell populations. Science 194(4260):23–28
20. Fang JS, Gillies RD, Gatenby RA (2008) Adaptation to hypoxia and acidosis in carcinogenesis and tumor progression. Semin Cancer Biol 18(5):330–337
21. Gatenby RA, Gillies RJ (2008) A microenvironmental model of carcinogenesis. Nat Rev Cancer 8(1):56–61
22. Graeber TG, Osmanian C, Jacks T, Housman DE, Koch CJ, Lowe SW, Giaccia AJ (1996) Hypoxia-mediated selection of cells with diminished apoptotic potential in solid tumours. Nature 379(6560):88–91
23. Yun J, Rago C, Cheong I, Pagliarini R, Angenendt P, Rajagopalan H, Schmidt K, Willson JK, Markowitz S, Zhou S et al (2009) Glucose deprivation contributes to the development of KRAS pathway mutations in tumor cells. Science 325(5947):1555–1559
24. Flier JS, Mueckler MM, Usher P, Lodish HF (1987) Elevated levels of glucose transport and transporter messenger RNA are induced by ras or src oncogenes. Science 235(4795):1492–1495
25. Coquelle A, Toledo F, Stern S, Bieth A, Debatisse M (1998) A new role for hypoxia in tumor progression: induction of fragile site triggering genomic rearrangements and formation of complex DMs and HSRs. Mol Cell 2(2):259–265
26. Yuan J, Narayanan L, Rockwell S, Glazer PM (2000) Diminished DNA repair and elevated mutagenesis in mammalian cells exposed to hypoxia and low pH. Cancer Res 60(16):4372–4376
27. Rice GC, Hoy C, Schimke RT (1986) Transient hypoxia enhances the frequency of dihydrofolate reductase gene amplification in Chinese hamster ovary cells. Proc Natl Acad Sci U S A 83(16):5978–5982
28. Basanta D, Ribba B, Watkin E, You B, Deutsch A (2011) Computational analysis of the influence of the microenvironment on carcinogenesis. Math Biosci 229(1):22–29
29. Little MP (2010) Cancer models, genomic instability and somatic cellular Darwinian evolution. Biol Direct 5:19 (discussion 19)
30. Gatenby RA, Brown J, Vincent T (2009) Lessons from applied ecology: cancer control using an evolutionary double bind. Cancer Res 69(19):7499–7502
31. Anderson AR, Hassanein M, Branch KM, Lu J, Lobdell NA, Maier J, Basanta D, Weidow B, Narasanna A, Arteaga CL et al (2009) Microenvironmental independence associated with tumor progression. Cancer Res 69(22):8797–8806
32. Tang X, Lucas JE, Chen JL, LaMonte G, Wu J, Wang MC, Koumenis C, Chi JT (2012) Functional interaction between responses to lactic acidosis and hypoxia regulates genomic transcriptional outputs. Cancer Res 72(2):491–502
33. Lucas JE, Kung HN, Chi JT (2010) Latent factor analysis to discover pathway-associated putative segmental aneuploidies in human cancers. PLoS Comput Biol 6(9):e1000920

Part I
“Sensing Acidity”

Chapter 2
The Molecular Mechanism of Cellular Sensing of Acidity

Zaven O'Bryant and Zhigang Xiong

Introduction

Acidic conditions often accompany a variety of pathologic conditions such as inflammation and ischemia. They may be caused by defective pH stabilization mechanism that is usually the result of tissue injury, increased metabolism, hypoxia, hypercapnia, and increased activity. Inability to maintain pH during stroke, for example, leads to worse prognosis [66, 67]. On the other hand, acidosis activates sensory neurons in the case of myocardial infarct and tissue inflammation. Examples have been shown where animals have even utilized toxins, to enhance or protect against acidic conditions [10, 21, 25, 50]. Cellular sensing and appropriate response to acid by ion channels and receptors to extracellular pH (pH_0) or intracellular pH (pH_i) is necessary for organ systems to maintain proper physiologic conditions.

Various microenvironments such as ones found in brain are particularly sensitive to pH changes which may exert pathologic downstream consequences. Likewise, cells in nonneuronal microenvironments also maintain physiologic proton gradients that vary between the extracellular milieu and the cytosol. pH_0 varies in each anatomical compartment with normal pH_0 typically between 7.3 and 7.4; both nonneuronal and neuronal cells alike have the capacity to maintain intracellular pH between ~7.3 and ~7.0 [66]. However, in pathologic conditions like ischemic brain, pH_0 may fall to 6.5–6.0 [66].

Change in pH_0 or pH_i directly affects various membrane receptors and transporters where activation or inhibition of these molecules may be initiated by pH alterations. Acid-sensing ions channels (ASICs), identified as a subfamily of the epithelial sodium channel/degenerin family (DEG/ENaC), are directly activated by an increase in extracellular proton concentration [59]. Found in the central nervous system (CNS) and peripheral nervous system (PNS), sensing pH_0 changes by ASICs contributes to CNS neuronal function and higher-order processes like learning and memory, synaptic plasticity, anxiety, depression, and seizure termination. ASICs

Z. Xiong (✉) · Z. O'Bryant
Department of Neurobiology, Morehouse School of Medicine, Atlanta, GA, USA
e-mail: zxiong@msm.edu

J-T. A. Chi (ed.), *Molecular Genetics of Dysregulated pH Homeostasis*,
DOI 10.1007/978-1-4939-1683-2_2

participate in sensory transduction, mechanosensation, pain, retinal function and other roles have also been described [1, 38, 52, 63]. ASICs have been the subject of increasing interest due to their direct relationship with acidosis-related neuronal injury and have provided a promising new target for therapeutic intervention of stroke. In contrast to ASICs, several voltage-gated and ligand-gated channels are inhibited during acidic conditions. For example, N-methyl-D-aspartate (NMDA) channels are ligand-gated channels that are inhibited by a decrease in pH_o [27]. Similar to ASICs, some members of the transient receptor potential (TRP) family of channels have also been shown to have pH_o sensitivity where a decrease in pH_o causes channel activation [33]. The TRP superfamily was originally identified in *Drosophila* and is responsible for sensory transduction of the retina [20]. The TRP family is found in excitable and nonexcitable cells and is responsible for a wide variety of pathological conditions [30]. Specific receptors and transporters can also be influenced by changes in pH_i. ASIC activation and inactivation, for example, can be modulated by pH_i; such changes can be caused by drugs or endogenous mechanisms [60].

This chapter focuses on the cellular and molecular mechanisms of sensing acidity of ASICs primarily in the nervous system. However, nonneuronal mechanisms and other ion channels/receptors to sense acidic conditions are also addressed.

Biochemistry of Acid-Sensing Ions Channels (ASICs) and Other Acid-Sensing Molecules

ASICs are voltage-independent, amiloride-sensitive, cation-selective ion channels which belong to the DEG/ENaC family of channels [52]. There are four genes (ASIC 1–4) in mammals that encode six distinct subunits ASIC1a, ASIC1b, ASIC2a, ASIC2b, ASIC3, and ASIC4 [67]. The sub-distinction designation of "a" or "b" corresponds to the differences between the NH_2 termini and location of the alternate splice site for ASIC1 and ASIC2. Recently resolved crystal structure from chicken, *Gallus gallus,* shows that ASIC subunits assemble as trimers to form functional channels which can be homomeric, consisting of identical subunits, or heteromeric, consisting of different subunits [31, 35]. Caveats to homomeric formation, ASIC2b and ASIC4 do not form functional channels but may associate with other subunits to form functional heteromeric channels. The formation of heteromers along with subunit combination and stoichiometry dictate channel properties such as gating, permeability, and activation. While ASIC1a homomers, which are ubiquitously expressed in the nervous system, respond to low pH_o by mediating a fast and transient inward current with a threshold pH of 7.0 and the pH for half maximal activation ($pH_{0.5}$) at 6.2 [67], homomeric ASIC3 channels, mostly found in the PNS, respond to pH_o drops by a biphasic response with a fast desensitizing current followed by a sustained component and a high sensitivity to protons with a $pH_{0.5}$ of 6.7 [63].

Each ASIC subunit consists of two transmembrane domains, TM1 and TM2, and a large cystine-rich extracellular loop. Both NH_2 and COOH termini of ASICs lie within the intracellular space [52]. The extracellular domain (ED) of ASIC1a consists of 318 of 528 total residues with many crevices and intersubunit contact sites which have a role in gating or binding of other modulators of channel activity such as zinc [18]. Heteromeric and homomeric ASIC1b is one example where Zn^{2+} regulates channel activity by acting on cysteine149 located in the ED [36]. Other metals such as Cu^{2+} inhibits the ASICs and reduces acid-mediated membrane depolarization [61]. Reduction in extracellular $[Ca^{2+}]$ is known to regulate the activities of ASIC1a, ASIC1b, and ASIC3 channels as well as subfamily member of TRP channels, e.g., TRPM7 [20, 47].

ASIC Structure and Function

ASICs are closed when extracellular protons have not reached the channels' respective threshold concentration. When bound by protons, channels rapidly activate and then desensitize [57]. ASICs reside in at least three conformational states, the first is a nonconducting resting state where the channel can be activated in response to decrease in pH, a second conducting or open state where ions are transported through the pore and lastly, a desensitized or nonconducting state where $[pH_o]$ is not able to cause further activation [46]. The ED of ASIC1a protrudes from the plane of the membrane outward and is organized into discrete regions. Using the analogy of a hand, ASIC regions have been named "palm, knuckle, β-ball, finger, and thumb" domains based on their biochemical arrangement [35, 49, 72]. Residues within TM2 constitute the pore while TM1 also contributes a segment in the extracellular vestibule of the pore [31, 52]. The pore of ASIC1a forms an hourglass configuration in the desensitized state with contribution from TM2 which holds the residues in place [35, 57]. In addition, TM2 helices are tilted at a 50° angle from the membrane and the intersection with TM1 residues from the outer vestibule are speculated to form a desensitization gate [13, 31, 57]. A cluster of negatively charged residues have been suggested to form the tentative proton sensor in the cleft between the "thumb" and "finger" domains of ASIC1, in which carboxyl–carboxylate interaction pairs may play a role in channel gating [35]. Two carboxyl–carboxylate pairs E230 through D358 and E212 through D414 and a carboxyl–hydroxyl interaction pair E23 through S354 contribute to the underlying structure [51]. The aforementioned region can be referred to as the acidic pocket and contributes to a highly negative potential serving as binding sites for Ca^{2+}, Na^{2+} and H^+ [51]. The binding of either of the cations affects H^+ binding and vice versa [51]. Recently, a series of publications have elucidated the intimate structure/function relationship [5, 7, 39]. X-ray crystallography has determined that the TM2 can be divided into two distinct regions, TM2a and TM2b, and that the break in the helical structure is approximately parallel to the membrane [5]. Interestingly, the TM2b element interacts with the TM1 region of the neighboring subunit generating a continuous TM2b helical segment [6]. This geometry allows the area between the two TM segments, termed the "GAS" (glycine, alanine, serine) selectivity filter, to effectively form a GAS belt [5]. From this arrangement, the GAS filter helps to form a component of the ion channel pore [6]. The entire structure of the pore consists of a gate and several integrated vestibules in a similar fashion to P2x receptors [7].

ASIC3 characteristically has a biphasic current in response to acidosis with a rapidly inactivating peak current and sustained plateau phase that persists as long as there is acid stimulation [23]. This is in contrast to the ASIC1a current profile which displays a rapidly inactivating transient current that deactivates in the presence of acid. As member of the Na^+ family of channels, ASICs have a reversal potential near +60 mV, so that when activated at typical resting potentials causes membrane depolarization. Data from several studies suggest that this depolarization is tied to Ca^{2+} influx causing neuronal injury [60, 68, 73].

Functional expression of ASIC channels relies on several factors including the trafficking of the molecule to the cell surface. Donier et al. reported that annexin light chain p11, a member of the S100 small phospholipid and Ca^{2+}-binding protein family, utilizes the intracellular N-terminus of ASIC1a in a heterologous system to traffic the protein to the membrane [26]. ASIC trafficking also requires PICK1, protein interacting with C-kinase-1, which regulates membrane incorporation and function of ASIC1a in a lipid binding-dependent manner [37].

Gating, Permeability, and Proton Interactions

ASIC1a homomeric channels have clear permeability to Ca^{2+}, establishing them as another vanguard of Ca^{2+}-mediated ischemic brain injury [41, 67, 73]. Despite the ability to conduct Ca^{2+} ions, there are data suggesting that external increased [Ca^{2+}] can reduce the conductance of ASIC1a and 2a [22]. Notably, the hallmark of ASICs is the characteristic permeability to Na^+ ions. Compared to other Na^+, ASICs are less permeable to K^+, Li^+, and H^+ ions; additionally, subunit composition greatly affects the permeability of heterologous systems like ASIC1a+ASIC2. Interestingly, ASICs are inhibited by heavy metals Gd^{3+}, Pb^{2+}, Ni^{2+}, Cd^{2+}, Cu^{2+}, and Mg^2. Zn^{2+} conversely has a dual effect, potentiating at micromolar concentrations and inhibiting at nanomolar concentrations [18]. Interestingly, openings of chloride channels (ClC), for example, by γ-aminobutyric acid (GABA) or glycine, modulate the ASIC activity in neurons [17].

There is a wealth of information supporting that activation of ASICs requires protonation of multiple residues [31, 42, 46, 57]. The pre-TM2 region is essential for gating and ion permeability and changes in gating are speculated to be done through the reorganization of the palm domain in the extracellular region. For example, interplay between ASIC1a subunits requires that the opening of the conductive pathway for ion flow involves the coordinated rotation of TM2 [46, 72]. ASIC1 proton sensitivity is conferred by amino acid substitutions Q77L and

T85L of the external segment of TM1 [32]. The arrangement between TM1 and TM2 is such that TM1 forms the exterior of the TM complexes mostly in contact with the lipid bilayer while TM2 forms the pore complex [31]. Four vestibules organize the passage of ions through the ASIC protein complex. Fenestrations before the extracellular vestibule are located just prior to the pore and proximal to the wrist region. Symmetrically related carboxyl moieties contribute to pore occlusion and concentration of cations leads to increased channel conductance [31, 57]. Located halfway between the lipid bilayer, the pore forms the desensitization gate by a constriction of the TM2 complexes and several key residues conferring physical blockage of the passage after extending activation by H^+ [31]. However, to undergo the physical blockade, only a small conformational change is required by the selectivity filter and the spatially distinct ion channel gate [5].

The interface between the binding and pore domains plays a critical role in the gating of Cys loop ligand-activated receptors [13]. Protons bind just before TM1 and tryptophan (Trp) regions under the thumb where they enact proton-mediated conformational changes in the pre-TM2 region allowing cation flux through the pore [35]. A critical region of the thumb associates with chloride ions suggesting that there is an interaction or relationship between ASICs and ligand-gated CIC [17]. Another region of proton binding is adjacent to the extracellular vestibule. In all, three binding sites may be necessary in the open, conducting state of the channel [31]. Proton binding sites and conformational changes in the TM domains imply that the trigonal antiprism coordination is the optimal arrangement of proton binding. Structurally, the N-terminal of the TM1 region may stabilize the channel in the open state [13].

It has been suggested that ASICs can be activated by nonproton ligands as well, and the structure of the ED may contribute functionally to the protein [42]. 2-guanidine-4-methylquinazoline (GMQ) is a guanidinium heterocyclic ring and an example of a nonproton ligand for ASIC3 which interacts with the subunit at ED E79 and E423 [74]. Later, the same group was able to elucidate the binding mode and architecture of the nonproton binding domains of GMQ required for ASIC3 activation [75]. Screening approaches for alternative ASIC channel ligands also identified Phe-Met-Arg-Phe (FMRF) amide and neuropeptide FF, found in humans, in modifying channel properties by potentiating or inducing a sustained component but not activation of ASICs [2]. Other FMRF-like peptides, dynorphin A and big dynorphin, reduce steady-state inactivation of the ASIC1a channels [19]. Mentioned previously, animal toxins and venoms like MitTx, isolated from the venom of the Texas coral snake *Micrurus tener tener,* cause pain by activation of ASIC1a and/or ASIC1b [10]. Psalmotoxin (PcTx1), isolated from the venom of the South American tarantula *Psalmopoeus cambridgei,* however inhibits ASIC1a channels by shifting the channels into the desensitized state [16]. Furthermore, other peptides and small molecules inhibit ASIC1, ASIC2, and ASIC3 currents such as mambalgin-1, isolated from the venom of the black mamba snake, *Dendroaspis polylepis polylepis,* A-317567, amiloride, and Sevanol, isolated from a plant *Thymus armeniacus* [76]. Pharmacological use of these compounds is being explored to exploit their therapeutic benefit for pain, stroke, and other neurologic diseases.

Other Acid-Sensing Molecules

Of the acid-sensing molecules, ASICs and TRPV1 are the most intensively studied. However, other acid-sensing members include potassium channels such as members of the two-pore-domain K^+ channels are differentially regulated by small deviations of extra- or intracellular pH from physiological levels [33, 33]. Other members of the TRP superfamily, TRPV4, TRPC4, TRPC5, TRPP2, along with ionotropic purinoceptors (P2X), inward rectifier K^+ channels, voltage-activated K^+ channels, L-type Ca^{2+} channels, hyperpolarization-activated cyclic nucleotide-gated channels, gap junction channels, and Cl^- channels display some acid sensitivity [33].

The superfamily of voltage-gated ClC is ubiquitously expressed in many tissues such as the brain, muscle, and kidney. ClC are activated, inhibited, or gated by pH changes [4]. Pharmacologic agents 4,4′-diisothiocyanatostilbene-2,2′-disulfonic acid (DIDS) and diphenylamine-2-carboxylic (DPC) acid inhibit activity. One such microenvironment within the testes is acidified by fluid secreted from Sertoli cells. The change in pH is then sensed by extracellular acidic pH-activated CIC (also named as acid-sensing Cl^- channel, ASCC). ASCCs elicits an outwardly rectifying current in Sertoli cells [4], human umbilical cord vein endothelia cells [43], lung epithelial cells [9], myocytes [71], and monocyte/macrophage [53]. Still, details of channel function and therapeutic potential have yet to be identified for this recent, yet little investigated ion channel.

Previously mentioned, members of the TRP family can be modulated and/or activated by acidic pH. An example in detection of noxious stimuli is the transient receptor potential vanilloid 1 (TRPV1) channel, or capsaicin receptor. Also found in neuronal and nonneuronal tissues [30, 77], TRPV1 is implicated in the development of diseases such as inflammatory heat hyperalgesia, visceral hyperreflexia, and airway inflammation among others [3]. The channel consists of six TM spanning segments with the pore located between TM5 and TM6 [20, 33]. Additionally, four subunits are assembled as tetramers to form a functional channel. Most TRPs behave as nonselective cation channels, with Ca^{2+} permeability which is a salient feature of TRPV1 conductance [33]. Likewise, transient receptor protein melastatin-7 (TRPM7) also conducts Ca^{2+} [58] and Zn^{2+} [34]. During ischemic conditions, intracellular [Ca^{2+}] increases leading to Ca^{2+} overload and cell toxicity [33, 34, 58]. In conditions such as diabetes where acidic conditions are associated with inflammation, Ca^{2+} influx via TRP channels can cause activation of monocytes [65], thus modulating the immune response.

Potassium channels such as ATP-sensitive K^+ channels (K_{ATP}) have also been shown to be sensitive to acid. Potassium inward rectifier 6.2 ($K_{ir}6.2$) subunits in combination with SUR1 in oocytes display a graded response to pH_o in both the presence or absence of CO_2, the response is modulated by pH_i manipulation [69].

Systems Biology

The Eye

The distribution of the ASICs is concentrated in the CNS/PNS. Sensory organs like the eye, where cells in the microenvironment of the retina express ASICs, are sensitive to pH fluctuation during ocular diseases [56]. Loss of retinal ganglion cells of the eye during ischemia contributes to poor vision and may eventually lead to blindness. Data have shown that in hypoxic conditions, Ca^{2+}-permeable ASIC1a channels enact cytotoxic injury through Ca^{2+}-mediated mechanisms leading to neuron death. So far, at least three studies have investigated ASIC expression and functionality in the eye [11, 44, 56].

Chemosensation

Other sensory modalities, like the tongue, have chemosensory neuronal innervations to convey taste information. ASICs have been implicated in taste sensation [54] and the superfamily of DEG/ENaC channels, in general, have been implicated in chemosensation transduction pathways involved in salt and sour taste [8, 14].

Synaptic Transmission

It is known that ASICs participate in synaptic transmission [62]. The hippocampus is an essential part of the brain that contributes to learning and memory. ASICs expressed in neurons found in the hippocampus and located at the postsynaptic membrane may be stimulated by ejection of protons from synaptic vesicles [62]. Mentioned earlier, ASICs are found in many brain regions. The hippocampus and cortex are principal areas being researched, where neurons can be isolated for culture and their characteristics pertaining to ASICs can be studied. At the cellular level, ASICs localize to the soma, the dendrites, and most importantly the synapse. The composition of ASIC expression along the neuron no doubt varies in the CNS, and neurons are most likely to express ASIC1a across all areas of the membrane. Loss of ASIC1a directly contributes to loss or impairment of spatial learning, fear and anxiety, eye blink conditioning, and other behaviors [62]. In a seminal paper, Wemmie et al. concluded that ASICs are expressed in synaptosomes and that the disruption of ASIC expression results in impaired hippocampal long-term potentiation (LTP)[62]. They also suggest that NMDA channels, also involved in learning and memory, and synaptic plasticity, are impacted by activation of ASIC channels at the postsynaptic membrane [62].

Cardiovascular/Respiratory

Physiologic mechanisms to detect and respond to pH changes are distributed widely. Examples of critical organs such as the central and peripheral chemoreceptors, aortic, and carotid bodies use pH sensing to maintain tissue homeostasis. Focusing on the centralized components, cardiovascular and respiratory centers are located in the brain within the pneumotaxic center, the dorsal respiratory group, and lastly the ventral respiratory group [45]. CO_2 rests in equilibrium with carbonic acid and when dissociated, separates into bicarbonate and H^+; the greater the amount of CO_2, the lower the pH. Increases in CO_2 are subsequently sensed as acid in the peripheral chemoreceptors and triggers afferent neuron excitation [45]. In turn, this signal is relayed to central receptors and areas of the brain that regulate breathing. Could ASICs or other mechanisms be involved in this process? More work is needed to clearly define how chemoreceptors process information on pH_o. In guinea pig, acidosis decreases the basal tone of tracheal rings. This acid-induced airway relaxation seems to be independent of sensory nerves, suggesting a regulation of airway basal tone mediated by smooth muscle ASICs [28].

The detection of ischemia-related cardiac acidosis is sensed by cellular mechanisms such as ASIC1a, ASIC3, TRPV-1, and outwardly rectifying Cl^- channels ($I_{Cl\ acid}$). Mentioned earlier, tissue acidosis is a major stimulus for afferent neuron signaling initiated via peripherally located ASICs. Found in mouse and guinea pig, atrial and ventricular cardiomyocytes, $I_{Cl\ acid}$ are stimulated by a decrease in pH_o causing an outward flux of anions [71]. This response is in line with tissue acidosis resultant of lactate but the physiological significance of this discovery has yet to be determined. More supporting in vitro evidence for acid sensing in cardiac dorsal root ganglion show that ASIC3-mediated current most clearly mimics in vivo physiology. A characteristic of ASIC3-mediated current is the sustained component which may contribute to the long-lasting sensation of angina [70]. However, ASIC current, also dependent on subunit composition, is modulated by anions affecting the pH dependence of activation and desensitization kinetics [40]. This might lay a possible link between anion flux by $I_{Cl\ acid}$, a possible extracellular fluid composition sensing mechanism of ASICs, along with pH_o sensing. Highlighting this proposed extracellular fluid sampling paradigm, in metabolic diseases such as diabetes mellitus, ASIC3's exquisite sensitivity to H^+ plays a role in insulin resistance by yet unresolved mechanisms [64]. The aforementioned ligand-gated ClC may play a role in regulating ASIC inhibition by concurrent opening of GABA receptors [17].

Pathologies Associated with Dysregulation of pH Sensing

CNS: Acidosis-Mediated, Glutamate-Independent Neuronal Injury

ASICs are largely responsible for acidosis-mediated, glutamate-independent neuronal injury in ischemic brain [48, 66, 68]. Injury is at least partially caused by a flux of Ca^{2+} that cannot be overcome by endoplasmic reticulum buffering. In addition, Ca^{2+} calmodulin-dependent protein kinase II (CaMKII) is also coupled with ASIC1a activation, promoting acidosis-mediated ischemic damage [29]. Trafficking and scaffold protein PICK1 also plays a role in ASIC1a-mediated neuronal injury by increasing surface expression of the ion channel possibly during acidic conditions [37].

Recently discovered mambalgin-1 attenuates pain in the PNS by limiting activation of ASIC1a, ASIC1b, ASIC1a+ASIC2b, and ASIC1a+ASIC2a homomers and heteromers, respectively [25]. Perhaps this newly derived peptide might protect against ASIC-mediated stroke pathologies primarily because ASIC1a+ASIC2a heteromers are highly expressed in the brain. Currently, protection from stroke using the nonspecific ASIC antagonist amiloride has proven successful in animal models but lacks the specificity of mambalgin-1 and PcTX1 [68]. Therapeutic cocktails of ASIC inhibitors may be prepared and tested, in combination, for additive or synergistic effect. Additionally, protection can be effected though gene knockout in experimental animals or directly by intracerbroventricular administration of $NaHCO_3^-$ [48].

PNS: Pain, Acidosis

In the PNS, sensory transduction is carried by neurons containing various ASIC subunits, some of which are homomeric and others form heteromultimeric channels. In addition to ASIC1a, ASIC2 splice variants 2a and 2b are expressed in the dorsal root ganglion [24] and colocalize with ASIC3 [1]. Alverez de la Rosa speculates that the ASIC2+ASIC3 pairing may serve as a mechanism for mechanotransduction activated by stretch, swelling, or suction applied to the cell membrane as found in *Caenorhabditis elegans* which contains an orthologous ASIC2 mechanosensation mechanism [1, 24]. Additionally, ASIC1 and ASIC3 homomers were identified in dorsal root ganglia (DRG) where ASIC3 contributes to pain transduction [1, 24, 55]. Inflammation may cause tissue acidosis leading to depolarization and sensation of pain. Conversely, experimental ASIC3 silencing has led to reduced pain sensitization in mice [23]. In humans, aberrant expression of ASIC3 could elicit pain from slight deviation in pH [23].

PNS transduction of pain is relayed from sensory nerve endings that are stimulated by heat, cold, mechanical deformation, inflammation, tissue damage, and chemical stimuli. ASIC1a and ASIC3 subunits are both expressed in the PNS and are the subject of intense investigation to find alternative therapeutic remedies for analgesia. $ASIC3^{-/-}$ knockout animals display a variety of physiological characteristics modulating pain perspective, especially at moderate and high-intensity pain [15]. Perhaps there is an indirect link between the activity of ASIC1a- and ASIC3-mediated pain during periods where high or chronic stress may prevail? Nevertheless, a complex interplay exists between ASIC subunits as exemplified by ASIC3/ASIC2b pairing while prevalent paradigms have suggested that ASIC3 homomers are the sole mediator of pain [24]. The interrelated nature of ASICs interaction has set forth a new postulate that a triple gene knockout of ASIC1a, 2 and 3 would have an effect on mechanosensation and pain. Interestingly, a total knockout of ASICs in mice (excluding ASIC4) did not decrease sensation of pain or decrease nerve firing; but, increased cutaneous mechanosensation [38]. Compensatory up- or downregulation of other ion channels might be involved in mechanosensation and detection of noxious stimuli like that of localized tissue acidosis.

Still further, other mechanisms may mediate pain due to acidosis. Recently, an interesting phenomenon was reported by Qiu et al. that serotonin (5-HT_2) enhanced ASIC current in DRG neurons via intracellular mechanisms, although the specific subunit composition is yet to be determined [49].

Degenerative diseases such as osteoarthritis have pain mediated by ASIC3 which can be attenuated by ASIC3 selective blocker APETx2. Interestingly, ASIC3 polymorphisms are involved in human insulin resistance which is related to glucose metabolism [64]. Although seemingly unrelated, ASIC3 expression in peripheral sensory neurons may include metaboreceptive capabilities corresponding to tissue acidosis.

Summary and Conclusion

Key points:

- Acid activates ASICs from a constitutively closed configuration to an active conducting state and finally a desensitized state where the molecule is insensitive to protons [12].
- Multiple sites on the ED of ASIC1a contribute to acid sensing.
- Pore opening requires the coordinated rotation of TM2 domains.
- Other ligands such as FMRF amide and neuropeptide FF are able to modulate ASICs. Perhaps other endogenous ligands can activate this ion channel.
- ASICs are largely responsible for acidosis-mediated, glutamate-independent neuronal injury in the ischemic brain through the flux of Ca^{2+}.
- Other ion channels and receptors are involved in acid sensing: TRPs, N-methyl-D-aspartate receptors, outwardly rectifying Cl^- channels and ATP-sensitive K^+ channels.

Current evidence supports the role of ASICs and other receptors in acid sensing, acid-induced pain, acid-induced injury, and acid-induced feedback of homeostatic mechanisms. Characteristically expressed in neurons of the CNS and PNS, acid-sensing molecules are distributed throughout the body in a variety of tissues. The structure of each acid-sensing molecule is unique and is intimately related to the function of the channel. Each acid-sensing molecule has a specialized function and contributes to the overall balance of pH sensing and pH regulation. As a channel directly activated by extracellular protons, ASIC research should be considered to be at the forefront of disease where acidic conditions are generated.

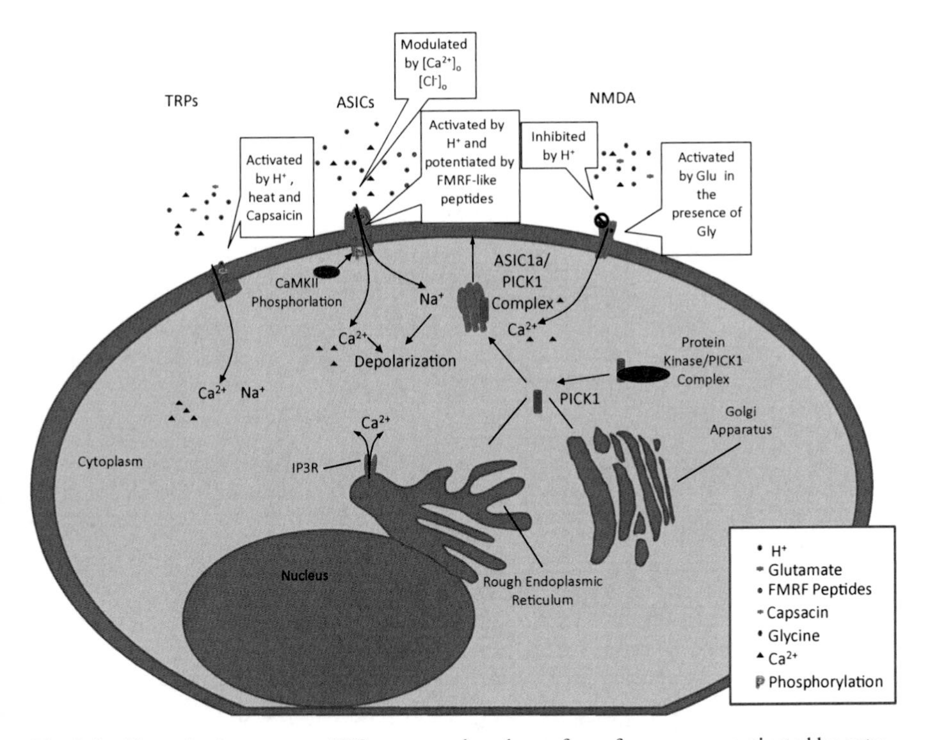

Fig. 2.1 pH sensing in neurons: *ASICs* expressed on the surface of neurons are activated by extracellular H^+. *ASIC1a* homomeric channels have clear permeability to Ca^{2+}, establishing them as mediators of Ca^{2+}-mediated ischemic neuronal injury. External $[Ca^{2+}]$ can reduce the conductance of *ASIC1a* and ASIC2a; there is speculation that ASIC3 becomes impermeable after high affinity Ca^{2+} binding. Phe-Met-Arg-Phe (*FMRF*) amide and neuropeptide FF, found in humans, modify channel properties by potentiating or inducing a sustained component but not activation of *ASICs*. N-methyl-D-aspartate (NMDA) channels, involved in learning, memory, and synaptic plasticity, are impacted by activation of *ASIC* channels due to membrane depolarization. Phosphorylation by *CaMKII* enhances *ASIC* activity as a result of NMDA activation. Protein interacting with C-kinase (*PICK1*) has been show to regulate the activity of *ASICs*. Transient receptor potential channels (*TRPs*) such as *TRPV1* can be activated by H^+ or capsaicin. Abbreviations: *ASICs* acid-sensing ion channels, *IP3R* inositol triphosphate receptor, *PICK1* protein interacting with C-kinase, *CaMKII* Ca^{2+}/calmodulin-dependent protein kinase II, *FMRF amide* Phe-Met-Arg-Phe neuropeptide

ASIC targeting may represent a new avenue for the treatment of injury caused by ischemic stroke and other neurological diseases. Other important channels like TRPV1 also represent a promising area of research related to pain and tissue inflammation. Furthermore, NMDA channels have also been studied intensively for their role in neuropathies but interestingly are linked to neuronal injury via functional coupling with ASICs [29]. In the PNS, new acid-sensing molecule antagonists are relevant to analgesic development for related diseases and conditions involving pain and inflammation. Please see Fig. 2.1. More studies are needed to further evaluate remaining mechanisms to sense acidity.

Acknowledgments Supported by NIH R01NS66027, NIMHD S21-MD000101, U54NS083932, AHA 0840132N, and ALZ IIRG-10–173350.

References

1. Alvarez de la Rosa D, Zhang P, Shao D, White F, Canessa CM (2002) Functional implications of the localization and activity of acid-sensitive channels in rat peripheral nervous system. Proc Natl Acad Sci U S A 99:2326–2331
2. Askwith CC, Cheng C, Ikuma M, Benson C, Price MP, Welsh MJ (2000) Neuropeptide FF and FMRFamide potentiate acid-evoked currents from sensory neurons and proton-gated DEG/ENaC channels. Neuron 26:133–141
3. Auzanneau C, Norez C, Antigny F, Thoreau V, Jougla C, Cantereau A, Becq F, Vandebrouck C (2007) Transient receptor potential vanilloid 1 (TRPV1) channels in cultured rat Sertoli cells regulate an acid sensing chloride channel. Biochem Pharmacol 75:476–483
4. Auzanneau C, Thoreau V, Kitzis A, Becq F (2003) A novel voltage-dependent chloride current activated by extracellular acidic pH in cultured rat Sertoli cells. J Biol Chem 278:19230–19236
5. Baconguis I, Bohlen C, Goehring A, Juilius D (2014) X-Ray structure of acid-sensing ion channel 1-snake toxin complex reveals open state of Na+. Cell 156:717–729
6. Baconguis I, Gouaux E (2012) Structural plasticity and dynamic selectivity of acid-sensing ion channel-spider toxin complexes. Nature 7416:400–405
7. Baconguis I, Hattori M, Gouaux E (2013) Unanticipated parallels in architecture and mechanism between ATP-gated P2X receptors and acid sending ion channels. Curr Opin Struct Biol 2:277–284
8. Ben-Shahar Y (2011) Sensory functions for degenerin/epithelial sodium channels (DEG/ENaC). Adv Genet 76:1–26
9. Blaisdell CJ, Edmonds RD, Wang XT, Guggino S, Zeitlin PL (2000) pH-regulated chloride secretion in fetal lung epithelia. Am J Physiol Lung Cell Mol Physiol 286:L1248–L1250
10. Bohlen CJ, Chesler AT, Sharif-Naeini R, Medzihradszky KF, Zhou S, King D, Sánchez EE, Burlingame AL, Basbaum AI, Julius D (2011) A heteromeric Texas coral snake toxin targets acid-sensing ion channels to produce pain. Nature 479:410–414
11. Brockway LM, Zhou ZH, Bubien JK, Jovov B, Benos DJ, Keyser KT (2002) Rabbit retinal neurons and glia express a variety of ENaC/DEG subunits. Am J Physiol Cell Physiol 283:C126–C134
12. Carattino MD (2011) Structural mechanisms underlying the function of epithelial sodium channel/acid-sensing ion channel. Curr Opin Nephrol Hypertens 20:555–560
13. Carattino MD, Della Vecchia MC (2012) Contribution of residues in second transmembrane domain of ASIC1a protein to ion selectivity. J Biol Chem 287:12927–93412
14. Chang RB, Waters H, Liman ER (2010) A proton current drives action potentials in genetically identified sour taste cells. Proc Natl Acad Sci U S A 107:22320–22325
15. Chen CC, Zimmer A, Sun WH, Hall J, Brownstein MJ (2002) A role for ASIC3 in the modulation of high-intensity pain stimuli. Proc Natl Acad Sci U S A 99:8992–8997
16. Chen X, Kalbacher H, Gründer S (2005) The tarantula toxin psalmotoxin 1 inhibits acid-sensing ion channel (ASIC) 1a by increasing its apparent H^+ affinity. J Gen Physiol 126:71–79
17. Chen X, Whissell P, Orser BA, MacDonald JF (2011) Functional modifications of acid-sensing ion channels by ligand-gated chloride channels. PLoS ONE 6:e21970
18. Chu XP, Wemmie JA, Wang WZ, Zhu XM, Saugstad JA, Price MP, Simon RP, Xiong ZG (2004) Subunit-dependent high-affinity zinc inhibition of acid-sensing ion channels. J Neurosci 24:8678–8689
19. Chu XP and Xiong ZG (2013) Acid-sensing ion channels in pathological conditions. Adv Exp Med Biol 961:419–431

20. Clapham DE, Montell C, Schultz G, Julius D (2002) The TRP Ion Channel Family. IUPHAR Compendium 2–40
21. Dawson RJ, Benz J, Stohler P, Tetaz T, Joseph C, Huber S, Schmid G, Hügin D, Pflimlin P, Trube G, Rudolph MG, Hennig M, Ruf A (2012) Structure of the acid-sensing ion channel 1 in complex with the gating modifier Psalmotoxin 1. Nat Commun 3:1–8
22. Deval E, Gasull X, Noël J, Salinas M, Baron A, Diochot S, Lingueglia E (2010) Acid-Sensing Ion Channels (ASICs): pharmacology and implication in pain. Pharmacol Ther 128:549–558
23. Deval E, Noël J, Lay N, Alloui A, Diochot S, Friend V, Jodar M, Lazdunski M, Lingueglia E (2008) ASIC3, a sensor of acidic and primary inflammatory pain. EMBO J 27:3047–3055
24. Deval E, Salinas M, Baron A, Lingueglia E, Lazdunski M (2004) ASIC2b-dependent regulation of ASIC3, an essential acid-sensing ion channel subunit in sensory neurons via the partner protein PICK-1. J Biol Chem 279:19531–19539
25. Diochot S, Baron A, Salinas M, Douguet D, Scarzello S, Dabert-Gay AS, Debayle D, Friend V, Alloui A, Lazdunski M, Lingueglia E (2012) Black mamba venom peptides target acid-sensing ion channels to abolish pain. Nature 490:552–555
26. Donier E, Rugiero F, Okuse K, Wood JN (2005) Annexin II light chain p11 promotes functional expression of acid-sensing ion channel ASIC1a. J Biol Chem 280:38666–38672
27. Dravid SM, Erreger K, Yuan H, Nicholson K, Le P, Lyuboslavsky P, Almonte A, Murray E, Mosely C, Barber J, French A, Balster R, Murray TF, Traynelis SF (2007) Subunit-specific mechanisms and proton sensitivity of NMDA receptor channel block. J Physiol 58:107–128
28. Faisy C, Planquette B, Naline E, Risse PA, Frossard, Fagon JY, Advenier C, Devillier P (2007) Acid-induced modulation of airway basal tone and contractility: role of acid-sensing ion channels (ASICs) and TRPV1 receptor. Life Sci 81:1094–1102
29. Gao J, Duan B, Wang DG, Deng XH, Zhang GY, Xu L, Xu TL (2005) Coupling between NMDA receptor and acid-sensing ion channel contributes to ischemic neuronal death. Neuron 48:635–646
30. Gohar O (2005) The transient receptor potential (TRP) ion channels. Modulator 20:20–23
31. Gonzales EB, Kawate T, Gouaux E (2009) Pore architecture and ion sites in acid-sensing ion channels and P2X receptors. Nature 460:599–604
32. Gründer S and Chen X (2010) Structure, function, pharmacology of acid-sensing ion channels (ASICs): focus on ASIC1a. Int J Physiol Pathophysiol Pharmacol 2:73–94
33. Holzer P (2009) Acid-sensitive ion channels and receptors. Handb Exp Pharmacol 194:283–332
34. Inoue K, Branigan D, Xiong ZG (2010) Zinc-induced neurotoxicity mediated by transient receptor potential melastatin 7 channels. J Biol Chem 285:7430–7439
35. Jasti J, Furukawa H, Gonzales EB, Gouaux E (2007) Structure of acid-sensing ion channel 1 at 1.9 A resolution and low pH. Nature 449:316–323
36. Jiang Q, Inoue K, Wu X, Papasian CJ, Wang JQ, Xiong ZG, Chu XP (2011) Cysteine 149 in the extracellular finger domain of acid-sensing ion channel 1b subunit is critical for zinc-mediated inhibition. Neuroscience 193:89–99
37. Jin W, Shen C, Jing L, Xia J. (2010) PICK1 regulates the trafficking of ASIC1a and acidotoxicity in a BAR domain lipid binding-dependent manner. Mol Brain 3:1–11
38. Kang S, Jang JH, Price MP, Gautam M, Benson CJ, Gong H, Welsh MJ, Brennan TJ (2012) Simultaneous disruption of mouse ASIC1a, ASIC2 and ASIC3 genes enhances cutaneous mechanosensitivity. PLoS One. 7(4):e35225
39. Krauson AJ, Rued AC, Carattino MD. (2013) Independent contribution of extracellular proton binding sites to ASIC1a activation. J Biol Chem 48:34375–34383
40. Kusama N, Gautam M, Harding AM, Snyder PM, Benson CJ (2013) Acid-sensing ion channels (ASICs) are differentially modulated by anions dependent on their subunit composition. Am J Physiol Cell Physiol 304:89–101
41. Li M, Inoue K, Branigan D, Kratzer E, Hansen JC, Chen JW, Simon RP, Xiong ZG (2010) Acid-sensing ion channels in acidosis-induced injury of human brain neurons. J Cereb Blood Flow Metab 30:1247–1260

42. Li WG, Yu Y, Huang C, Cao H, Xu TL (2012) Nonproton ligand sensing domain is required for paradoxical stimulation of acid-sensing ion channel 3 (ASIC3) channels by amiloride. J Biol Chem 286:42635–42646
43. Ma ZY, Zhang W, Chen L, Wang R, Kan XH, Sun GZ, Liu CX, Li L, Zhang Y (2008) A proton-activated, outwardly rectifying chloride channel in human umbilical vein endothelial cells. Biochem Biophys Res Commun 371:437–440
44. Maubaret C, Delettre C, Sola S, Hamel CP (2002) Identification of preferentially expressed mRNAs in retina and cochlea. DNA Cell Biol 21:781–791
45. Nattie E (1999) CO2, brainstem chemoreceptors and breathing. Prog Neurobiol 59:299–331
46. Passero CJ, Okumura S, Carattino MD (2009) Conformational changes associated with proton-dependent gating of ASIC1a. J Biol Chem 284:36473–36481
47. Paukert M, Babini E, Pusch M, Gründer S (2004) Identification of the Ca^{2+} blocking site of acid-sensing ion channel (ASIC) 1 implications for channel gating. J Gen Physiol 124:383–394
48. Pignataro G, Simon RP, Xiong ZG (2007a). Prolonged activation of ASIC1a and the time window for neuroprotection in cerebral ischaemia. Brain 130:151–158
49. Qiu F, Qiu CY, Liu YQ, Wu D, Li JD, Hu WP (2012) Potentiation of acid-sensing ion channel activity by the activation of 5-HT2 receptors in rat dorsal root ganglion neurons. Neuropharmacology 63:494–500
50. Salinas M, Rash LD, Baron A, Lambeau G, Escoubas P, Lazdunski M (2012) The receptor site of the spider toxin PcTx1 on the proton-gated cation channel ASIC1a. J Physiol 570:339–354
51. Shaikh SA and Tajkhorshid E (2008) Potential cation and H^+ binding sites in acid sensing ion channel-1. Biophys J 95:5153–5164
52. Sherwood TW, Frey EN, Askwith CC (2012) Structure and activity of the acid-sensing ion channels. Am J Physiol Cell Physiol 301:C699–C710
53. Shi CY, Wang R, Liu CX, Jiang H, Ma ZY, Li L, Zhang W (2009) Simvastatin inhibits acidic extracellular pH-activated, outward rectifying chloride currents in RAW264.7 monocytic-macrophage and human peripheral monocytes. Int Immunopharmacol 9:247–252
54. Shimada S, Ueda T, Ishida Y, Yamamoto T, Ugawa S (2006) Acid-sensing ion channels in taste buds. Arch Histol Cytol 69:227–231
55. Sluka KA, Price MP, Breese NM, Stucky CL, Wemmie JA, Welsh MJ (2003) Chronic hyperalgesia induced by repeated acid injections in muscle is abolished by the loss of ASIC3, but not ASIC1. Pain. 106:229–239
56. Tan J, Ye X, Xu Y, Wang H, Sheng M, Wang F (2011) Acid-sensing ion channel 1a is involved in retinal ganglion cell death induced by hypoxia. Mol Vis 17:3300–3308
57. Tolino LA, Okumura S, Kashlan OB, Carattino MD (2011) Insights into the mechanism of pore opening of acid-sensing ion channel 1a. J Biol Chem 286:16297–16307
58. Villena FN, Becerra A, Echeverryia C, Briceno N, Varela D, Sarmiento D, Simon RP (2011) Increased expression of the transient receptor potential melastatin 7 channel is critically involved in lipopolysaccharide-induced reactive oxygen species-mediated neuronal death. Antioxid Redox Signal 15:2425–2438
59. Waldmann R, Lazdunski M (1998) H^+-gated cation channels: neuronal acid sensors in the NaC/DEG family of ion channels. Curr Opin Neurobiol 8:418–424
60. Wang WZ, Chu XP, Li MH, Seeds J, Simon RP, Xiong ZG (2006) Modulation of acid-sensing ion channel currents, acid-induced increase of intracellular Ca^{2+}, and acidosis-mediated neuronal injury by intracellular pH. J Biol Chem 281:29369–29378
61. Wang W, Yu Y, Xu T-L (2007) Modulation of acid-sensing ion channels by Cu^{2+} in cultured hypothalamic neurons of the rat. Neuroscience 145:631–641
62. Wemmie JA, Chen J, Askwith CC, Hruska-Hageman AM, Price MP, Nolan BC, Yoder PG, Lamani E, Hoshi T, Freeman JH, Welsh MJ (2002) The acid-activated ion channel ASIC contributes to synaptic plasticity, learning, and memory. Neuron 34:463–477
63. Wemmie JA, Price MP, Welsh MJ (2006) Acid-sensing ion channels: advances, questions and therapeutic opportunities. TRENDS in Neurosciences 29:578–586
64. Wu S, Hsu LA, Chou HH, Teng MS, Chang HH, Yeh KH, Chen CC, Chang PY, Cheng CF, Ko YL (2013) Association between an ASIC3 gene variant and insulin resistance in Taiwanese. Clin Chim Acta 415:1132–1136

65. Wuensch T, Thilo F, Krueger K, Scholze A, Ristow M, Tepel M (2010) High glucose–induced oxidative stress increases transient receptor potential channel expression in human monocytes. Diabetes 59:844–849
66. Xiong ZG, Chu XP, Simon RP (2007) Acid sensing ion channels–novel therapeutic targets for ischemic brain injury. Front Biosci 12:1376–1386
67. Xiong ZG, Pignataro G, Li M, Chang SY, Simon RP (2008) Acid-sensing ion channels (ASICs) as pharmacological targets for neurodegenerative diseases. Curr Opin Pharmacol 8:25–32
68. Xiong ZG, Zhu XM, Chu XP, Minami M, Hey J, Wei WL, MacDonald JF, Wemmie JA, Price MP, Welsh MJ, Simon RP (2004) Neuroprotection in ischemia: blocking calcium-permeable acid-sensing ion channels. Cell 118:687–698
69. Xu H, Cui N, Yang Z, Wu J, Giwa LR, Abdulkadir L, Sharma P, Jiang C (2001) Direct activation of cloned K(atp) channels by intracellular acidosis. J Biol Chem 276:12898–12902
70. Yagi J, Wenk HN, Naves LA, McCleskey EW (2006) Sustained currents through ASIC3 ion channels at the modest pH changes that occur during myocardial ischemia. Circ Res 99:501–509
71. Yamamoto S, Ehara T (2005) Acidic extracellular pH-activated outwardly rectifying chloride current in mammalian cardiac myocytes. Am J Physiol Heart Circ Physiol 209:H1905-H1914
72. Yang H, Yu Y, Li WG, Yu F, Cao H, Xu TL, Jiang H (2009) Inherent dynamics of the acid-sensing ion channel 1 correlates with the gating mechanism. PLoS Biol 7:1–13
73. Yermolaieva O, Leonard AS, Schnizler MK, Abboud FM, Welsh MJ (2003) Extracellular acidosis increases neuronal cell calcium by activating acid-sensing ion channel 1a. Proc Natl Acad Sci U S A 101:6752–6757
74. Yu Y, Chen Z, Li WG, Cao H, Feng EG, Yu F, Liu H, Jiang H, Xu TL (2010) A nonproton ligand sensor in the acid-sensing ion channel. Neuron 68:61–72
75. Yu Y, Li WG, Chen Z, Cao H, Yang H, Jiang H, Xu TL (2011) Atomic level characterization of the nonproton ligand-sensing domain of ASIC3 channels. J Biol Chem 286:24996–25006
76. Zha XM (2013) Acid-sensing ion channels: trafficking and synaptic function. Mol Brain 6:1–28
77. Zhu Z, Luo Z, Ma S, Liu D (2011) TRP channels and their implications in metabolic diseases. Pflugers Arch—Eur J Physiol 461:211–223

Chapter 3
The Molecular Basis of Sour Sensing in Mammals

Jianghai Ho, Hiroaki Matsunami and Yoshiro Ishimaru

List of Abbreviations (In Alphabetical Order)

ASIC	Acid-sensing ion channel
Car4	Carbonic anhydrase 4
CN	Cranial nerve
CT	Chorda tympani
CvP	Circumvallate papillae
DEG/ENaC	*Caenorhabditis elegans* degenerin/human epithelium amiloride-sensitive Na^+ channel
DTA	Diphtheria toxin A fragment
FoP	Foliate papillae
FuP	Fungiform papillae
GAD	Glutamate decarboxylase
GG	Geniculate ganglion
GN	Glossopharyngeal
GPCR	G protein-coupled receptors
GSP	Greater superficial petrosal
HCN	Hyperpolarization-activated cyclic nucleotide-gated channel
K2P	Two-pore domain K^+ channels
NPG	Nodose/petrosal ganglion
NST	Nucleus of the solitary tract
PbN	Parabrachial nucleus
Pkd1l3	Polycystic kidney disease 1-like 3

H. Matsunami (✉) · J. Ho
Department of Molecular Genetics and Microbiology, Duke University Medical Center,
Box 3509, 264 CARL Bldg., 213 Research Dr., Durham, NC 27710, USA
e-mail: matsu004@mc.duke.edu

Department of Neurobiology, Duke University Medical Center,
Box 3209, 412 Research Drive, Durham, NC 27710, USA

Y. Ishimaru
Department of Applied Biological Chemistry, Graduate School of Agricultural and Life Sciences,
The University of Tokyo, 1–1-1 Yayoi, Bunkyo-ku, Tokyo 113–8657, Japan

J-T. A. Chi (ed.), *Molecular Genetics of Dysregulated pH Homeostasis*,
DOI 10.1007/978-1-4939-1683-2_3

Pkd2l1	Polycystic kidney disease 2-like 1
SP	Substance P
TRC	Taste receptor cell
TRPV1	Transient receptor potential, vanilloid receptor subtype-1
WGA	Wheat germ agglutinin

Introduction

Taste plays an important role for organisms in determining the properties of ingested substances by conveying important information on five basic taste modalities—sweet, salty, sour, bitter, and umami. Sweet, salty, and umami taste modalities convey the carbohydrate, electrolyte, and glutamate content of food, indicating its desirability and stimulating appetitive responses. Sour and bitter modalities, on the other hand, convey the presence of acidity and potential toxins, respectively, stimulating aversive responses to such tastes [1].

In recent years, the receptors mediating sweet, bitter, and umami tastes have been identified as members of the T1R and T2R G-protein-coupled receptor (GPCR) families, while the molecular mechanisms underlying sour taste have yet to be clearly elucidated [8, 27, 28, 33, 58].

Acidic Perceptions in Taste

Substances that stimulate acidic taste in the oral cavity can be divided into at least two categories. Mineral (strong) acids such as HCl are fully dissociated in aqueous solutions and are detected as sour tasting via protons (H^+, but more precisely hydronium, H_3O^+, ions). Organic (weak) acids like acetic acid (HOAc) on the other hand do not fully ionize in aqueous solutions, and in addition to forming protons (as with mineral acids), also exist as protonated (undissociated) species (HOAc) that appear to be able to diffuse directly into and acidify the cytosolic contents of taste receptor cells (TRCs) to stimulate sour taste [40]. This fits well with reported observations that organic acids are perceived to be more sour than mineral acids at the same pH [42], and that the sour taste threshold for HCl occurs at a lower pH than organic acids [16].

Interestingly, a study on the effects of taste adaptation and cross-adaptation to HCl and acetic acid in human subjects suggests the acid reception process following exposure to protons, and undissociated acids are different and perhaps independent [14], setting the stage for the possibility that there are multiple, even redundant, pathways that lead to sour TRC activation, which is substantiated by experimental evidence showing the involvement of both protons and undissociated acid molecules in the sour TRC activation process [7, 23]. This is echoed in the

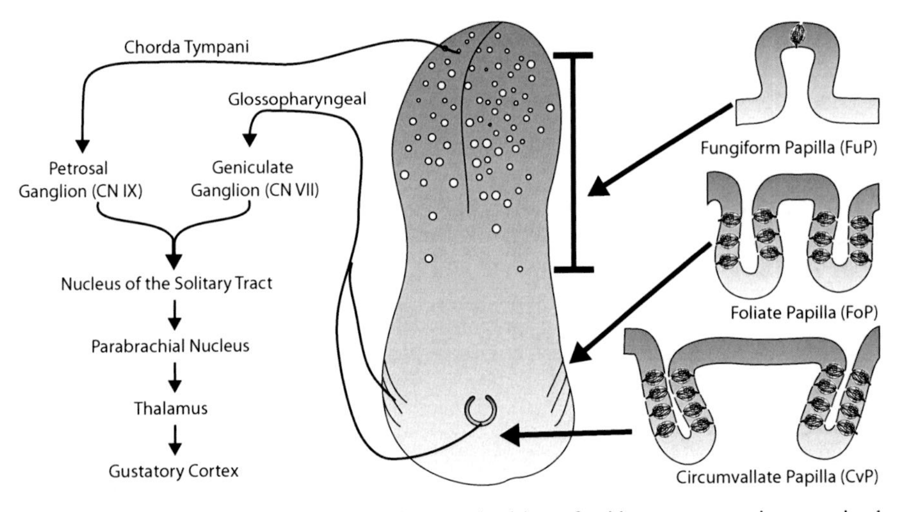

Fig. 3.1 Anatomy of taste. *Left panel*: The neural wiring of acid taste sensors is transmitted through the chorda tympani (CN IX) and glossopharyngeal (CN VII) nerves and converges in the NST and is transmitted to the gustatory cortex through the PbN and thalamus. *Center panel*: Diagrammatic representation of a rodent tongue, with FuP present in the anterior aspect, FoP present in anteriolateral folds, and a single CvP in the mid-anterior of the tongue. *Right panel*: Illustration of the shapes of different papillae and taste bud location

inability to predict the taste intensities of different sour-tasting compounds by pH (proton concentration) [42], titratable acidity (amount of NaOH required to raise the pH to 8.2) [61], or buffer capacity (amount of strong acid/base required to shift pH by 1 unit) [15]. More recently, it has also been proposed that the sour intensity perception is best predicted by the sum total of the concentrations of all the protonated molecular species plus the proton concentration [31], suggesting that both mechanisms of sour taste cell activation contribute to the differences in intensity of sour-tasting compounds.

Physiological Components of Taste

The Mammalian Taste System

Taste compounds taken into the oral cavity are detected by taste receptors localized in the taste pore areas at the apical end of taste buds [27, 28]. In the oral cavity, there are three major gustatory regions where taste buds are abundantly distributed: the circumvallate papillae (CvP), foliate papillae (FoP), and fungiform papillae (FuP), all found on the tongue, and also distributed singly on the soft palate ([47, 59, 60, 62, 67, 77]; Fig. 3.1, center and right panel).

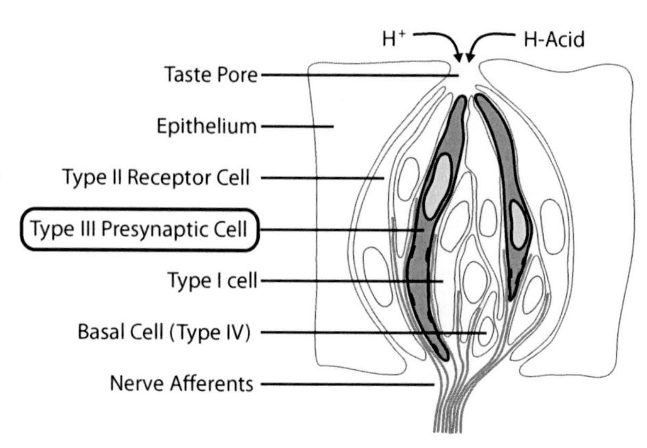

Fig. 3.2 Illustration of a taste bud. A taste bud is surrounded by lingual epithelia, and has a taste pore through which taste stimuli enter. There are four TRC cell types in the taste bud, type I glial-like cells, type II receptor cells, which detect sweet, umami, and bitter, type III presynaptic cells, which are the sour-transducing TRCs, and type IV basal cells thought to be progenitor cells responsible for replacing TRCs

Each taste bud forms an onion-like shape and is composed of 50–100 TRCs. Based on the intensity of staining and the ultrastructure of the cytoplasm, as observed by electron microscopy, the TRCs in each taste bud have been classified into four morphological types (Fig. 3.2). Type I (dark), type II (light), and type III (intermediate) taste cells are elongated and spindle shaped, whereas type IV taste cells are round; the latter are located at the bottom of the taste buds and are thought to be progenitor cells of other types of TRCs [8]. Recent studies revealed that each type of TRCs plays different roles in taste detection. Type I cells appear to be "glial-like" cells, morphologically wrapping around other TRCs and producing enzymes that degrade ATP released by type II cells [2]. Type II cells are responsible for the detection of sweet, bitter, or umami taste [10, 12]; they do not form conventional synapses and appear to release ATP as a transmitter, which appears to activate closely associated nerve afferents expressing P2X receptors [13, 76], along with adjacent taste cells expressing P2Y receptors [17, 24]. Type III cells that mediate sour taste transduction are on the other hand "presynaptic," [22, 23] forming synaptic contacts with the intragemmal nerve fibers and are thought to use serotonin (5-HT) as a neurotransmitter [24, 21].

Neural Wiring of Acid Sensing Systems

In general, taste buds are connected to the central nervous system (CNS) via cranial nerves (CN) VII (facial nerve) and IX (glossopharyngeal, GL, nerve; Fig. 3.1, left panel). Taste buds in the FuP are scattered on the anterior tongue, and together with those in the soft palate are innervated by the chorda tympani (CT) and the greater superficial petrosal (GSP) nerves. These nerves are branches of CN VII, and their cell bodies are accumulated in the geniculate ganglion (GG). In contrast, taste buds in the CvP and FoP are located in the posterior region of the tongue and are mostly innervated by CN IX, which has cell bodies in the nodose/petrosal ganglion (NPG). Taste information detected by taste buds at the periphery is transmitted to peripheral gustatory neurons in the geniculate, petrosal, and nodose ganglia and further conveyed to central gustatory neurons, including the nucleus of the solitary tract (NST), parabrachial nucleus (PbN), thalamus, and primary gustatory cortex in the insula [47, 59, 60, 62, 67, 77].

The wheat germ agglutinin (WGA) transgene has been used as a transneuronal tracer to label neural pathways that originate from cells genetically modified to express WGA in a variety of nervous systems, including the gustatory system [11, 50, 51, 78, 79]. To visualize the gustatory pathway that originates from sour-sensing TRCs in the posterior region of the tongue, transgenic mouse lines in which WGA was expressed in the type III taste cells of the CvP and FoP under the control of mouse polycystic kidney disease 1-like 3 (Pkd1l3) gene promoter/enhancer have been established [73]. Pkd1l3 exhibits specific expression in the TRCs of the CvP and FoP but not in the FuP [22, 29] or the solitary chemoreceptor cells of the nasal epithelium [50]. Pkd1l3-driven WGA not only confirmed innervation of the CvP and FoP by neurons of CN IX in the NPG but also revealed a small number innervated by neurons of CN VII in the GG via the CT, verifying previous electrophysiological reports of these connections in rats [74]. The transgenic mice genetically revealed the sour gustatory pathway from the Pkd1l3-positive sour TRCs in the posterior region of the tongue primarily through the peripheral gustatory neurons in the NPG and GG to the central neurons in the NST. This pathway was labeled separately from the trigeminal neural pathway, which mediates nociception.

Acid Nociception

In addition to the sour taste, acid at sufficient concentrations can also induce pain via nociceptors. These acid nociceptors have been extensively studied in the skin, and can be blocked by the general DEG/ENaC (*Caenorhabditis elegans* degenerin/human epithelium amiloride-sensitive Na+ channel) inhibitor amiloride. In the oral cavity, substance P (SP) expressing nerve fibers, which have been implicated in pain sensing in the peripheral nervous system (PNS) [43], have been found to innervate CvP and FuP via CN IX and V (trigeminal), respectively [18, 46], suggesting that these fibers could be involved in acid nociception in the oral cavity. pH-sensitive transient receptor potential, vanilloid receptor subtype-1 (TRPV1) and acid-sensitive ion channels (ASICs) are also expressed in human trigeminal ganglion neurons [70], which relay all nociceptive and somatosensory stimuli from the face and mouth to the CNS.

Molecular Basis of Acid Sensing

There have been a number of attempts at identifying the physiological acid sensor and signal transduction pathway that mediates acid sensing in taste cells, but thus far there have been no conclusive findings that point to a master acid sensor in mammalian TRCs. The search for this sensor has been complicated by the fact that virtually every protein contains amino acid residues that can bind protons, resulting in pH-dependent effects on most channels, transporters, and signal transduction molecules, further highlighting the importance of maintaining appropriate acid–base balance in most cells. This results in a level of difficulty to show that the pH effects on a specific protein are related to function of the physiological acid sensor without requiring sophisticated and complicated *in vivo* experiments.

While there is widespread acidification of TRCs on exposure, only a subpopulation of TRCs showed stimulus-related Ca^{2+} signals upon application of acid stimulus [55], which are mediated by action potential generation, leading to membrane depolarization and subsequent activation of voltage-gated Ca^{2+} channels, which could arise from reduced cytosolic pH buffer capacity or expression of proton-sensitive transduction molecules. These cells were subsequently identified as the presynaptic type III cells in taste buds [44], and intracellular protons have been proposed to be the proximate stimulus for acid sensing [23, 31].

In particular, the ability of both protons and protonated organic acids to mediate activation of sour sensing TRCs by seemingly separate pathways introduces two separate but related stimuli that contribute to the same sensory outcome, both of which require investigation. Further complicating the picture is the apparent activation of both bitter and sour TRCs by high concentrations of salt [52], which could mean that either the local pH around sour TRCs or their sensory mechanism is affected by salt content of the stimulus.

Perhaps because of these confounding factors, a large number of candidate molecules that could figure into the molecular basis of acid sensing have been identified by a number of laboratories, the most significant of which are Pkd1l3 and/or polycystic kidney disease 2-like 1 (Pkd2l1), ASICs, hyperpolarization-activated cyclic nucleotide-gated channels (HCNs), two-pore domain K^+ channels (K_2Ps), and carbonic anhydrase 4 (Car4; Fig. 3.3). Due to the many confounding factors previously explored, two main criteria must be met for the acid sensor to be identified—the molecule must be expressed in the appropriate (type III) TRCs, and it must be shown that an absence of the molecular sour-sensing TRCs in vivo eliminates or greatly diminishes the sour taste modality, and has no major effect on the other modalities. Currently, only Pkd2l1 [19] and Car4 [6] have been demonstrated to be required for sour taste detection in vivo by using gene knockout mice.

Pkd2l1/Pkd1l3

Pkd2l1 is a member of the TRP channel superfamily, and has six transmembrane (TM) domains and a putative pore region between the fifth and sixth TM domains (TM5 and TM6), similar to other TRP channel family members, most of which play important roles in signal transduction in various sensory systems, including vision, smell, pheromone, hearing, touch, osmolarity, thermosensation, and taste [53]. Pkd1l3 is a large protein with a very long N-terminal extracellular domain,

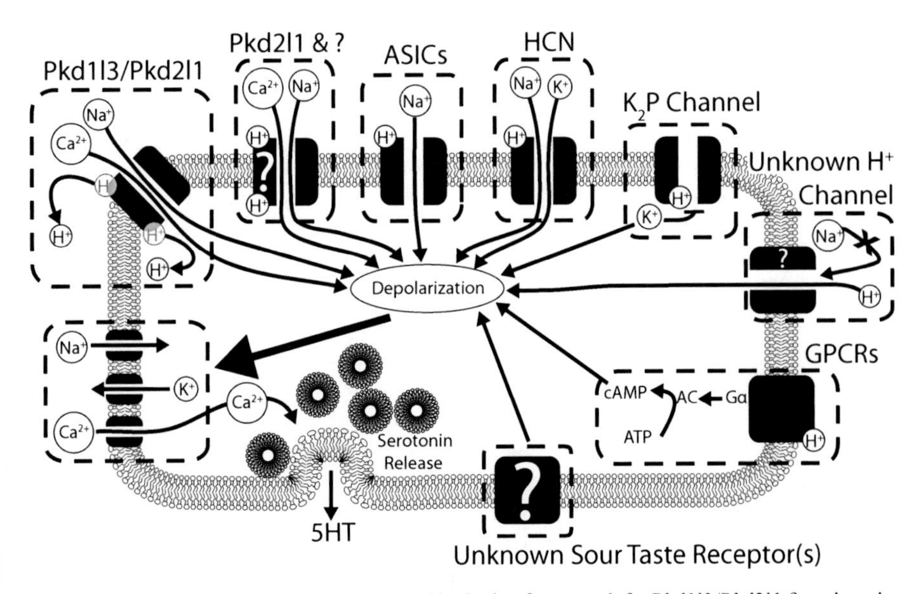

Fig. 3.3 Proposed sour sensing molecules. Clockwise from top left. *Pkd1l3/Pkd2l1* functions in CvP and FoP cells as an "off sensor" to signal the removal of an existing acid stimulus. *Pkd2l1* (along with an unknown partner) contributes to the sour-mediated depolarization of FuP TRCs. For both cases, it is likely that the conductances are similar to other TRP channels (Ca^{2+} and Na^{+}), but it is not clear whether intracellular or extracellular H^{+} gates the channel. *ASICs* are proposed to open in response to extracellular H^{+} and allow Na^{+} entry for TRC depolarization. *HCN channels* are also proposed to be gated by extracellular H^{+}, and permit Na^{+} and K^{+} movement. Intracellular H^{+} prevent K^{+} from exiting through K_2P channels, stopping the K^{+} leak current and depolarizing the TRC. A yet unknown H^{+} channel that is Na^{+} impermeable allows direct H^{+} entry into the cell for depolarization. Acid-sensitive *GPCRs* that function in kidney cells could mediate cell depolarization via the Gs-cyclic-AMP pathway. There are also likely additional unknown sour taste receptor(s) that have yet to be discovered. All of these mechanisms lead to generation of action potentials in the TRC, which activate voltage-gated Na^{+}, K^{+}, and Ca^{2+} channels and cause synaptic release of serotonin

followed by 11 TM domains that include a 6-TM TRP-like channel domain at the C-terminus. Due to structural differences, Pkd1l3 is not classified as a TRP channel, but is distantly related to TRP channels in amino acid sequence.

Pkd1l3 and Pkd2l1 are robustly coexpressed in the same subset of type III TRCs of the CvP and FoP, which are distinct from the sweet-, bitter-, and umami-sensing type II cells [22, 29]. In addition, Pkd2l1, but not Pkd1l3, is expressed in TRCs of the FuP and palate [22].

To investigate the role of Pkd2l1-expressing cells in mouse, transgenic mice in which Pkd2l1-expressing cells were genetically ablated using the diphtheria toxin A fragment (DTA) were generated [22]. In these animals, electrophysiological recordings of CT nerves revealed a complete lack of responses to sour stimuli, while a variety of taste stimuli representing the other taste modalities of sweet, salty, umami, and bitter were unaffected, indicating that these Pkd2l1-expressing type III cells function as sour TRCs.

Studies in a heterologous expression system using human embryonic kidney (HEK) 293T cells showed the interaction of PKD1L3 and PKD2L1 proteins with each other through their TM domains, and that this interaction is required for their functional cell surface expression [29, 30]. In addition, PKD2L1 protein did not localize to the taste pore but was distributed throughout the cytoplasm in TRCs of CvP and FoP in Pkd1l3 KO mice [30]. Functional analyses using Ca^{2+}-imaging and patch-clamp recordings also showed that HEK293T cells transfected with Pkd1l3 and Pkd2l1 specifically responded to a variety of sour compounds, including citric acid, hydrochloric acid, malic acid, and acetic acid, whereas they did not respond to other taste–quality classes [29].

Pkd1l3 and/or Pkd2l1 knockout mice were generated and analyzed using taste nerve recordings. In Pkd2l1 knockout mice, CT nerve responses to sour stimuli were significantly reduced compared with wild-type mice [19], even though Pkd1l3 is not expressed in the areas innervated by CT. In contrast, CT nerve responses in Pkd1l3 knockout mice and GL nerve responses in single- and double-KO (both Pkd1l3 and Pkd2l1) mice were comparable to the responses in wild-type mice [19, 48]. These results indicate that Pkd1l3 is not required for acid detection in TRCs, Pkd2l1 is required at least in part for acid sensing in TRCs innervated by the CT nerve (FuP and palate TRCs), and that there is a different sour-sensing mechanism for TRCs innervated predominantly by the GL nerve (FoP and CvP TRCs).

Subsequent studies revealed that the Pkd1l3/Pkd2l1 channel has a unique property, which we referred to as an "off-response," meaning that this channel is activated after the removal of an acid stimulus, even though initial acid exposure is essential [26]. In addition, Ca^{2+}-imaging and patch-clamp recordings using native taste cells revealed that off-responses upon acid stimulation were observed in type III cells isolated from the CvP (which coexpress Pkd1l3 and Pkd2l1) but not in cells isolated from the FuP (which expresses Pkd2l1 but not Pkd1l3) of glutamate decarboxylase (GAD) 67-green fluorescent protein (GFP) knock-in mice, which identify type III cells with GFP [32]. A similar analysis using taste cells isolated from Pkd1l3 and/or Pkd2l1 knockout mice may clarify whether and how these two molecules play crucial roles in the "off-response" in taste cells. Future studies will be needed to identify on-response receptors other than PKDs that are involved in sour taste detection, and the physiological relevance of the Pkd1l3/Pkd2l1-mediated off-response.

Acid Sensing Ion Channels

ASICs are proton-gated, amiloride-sensitive, voltage-insensitive cation channels belonging to the DEG/ENaC superfamily of ion channels [4]. There are at least six known mammalian ASIC subunits transcribled via alternate splicing from four ASIC loci (ASIC1–4 [37]), each containing two TM domains, a large extracellular loop and small intracellular loops forming homo- or heteromultimeric ion channels, which seem to favor Na^+ permeability when open [3]. ASIC expression seems to be enriched in neuronal tissues [38, 72], and in particular many subunits are expressed in dorsal root ganglion (DRG) nociceptive neurons [71], implying a role in pain perception at the periphery, and have been proposed to mediate proton-gated currents observed during myocardial ischemia [65, 69].

Therefore, for the obvious reason that ASICs are proton-gated ion channels, ASICs have been an obvious candidate as the acid sensor in TRCs. Early electrophysiological experiments showed that amiloride-sensitive channels are involved in sour taste in rats [49], suggesting the involvement of DEG/ENaC channels in sour taste sensation. Using a combination of homology screening and functional expression approaches, Ugawa et al. screened a rat CvP complementary DNA (cDNA) library and showed that ASIC2a and ASIC2b transcripts are present in taste cells, and ASIC2a/2b heteromeric channels are present in a subset of taste cells that resembled type III cells [68]. Unfortunately, ASIC2 subunits do not appear to be expressed in mouse taste cells, and ASIC2 KO mice show normal physiological responses to acid tastants [57]. ASIC1 subunits have been observed in mouse tongue [4], although there have been no reports examining its function in sour taste. Incidentally, ASIC3 is expressed in small diameter DRG neurons that coexpress SP in the PNS [54]. As SP-expressing neurons have been observed to innervate taste buds [18, 46], ASIC3 may play a role in the nociceptive sensing of acid, which has yet to be fully explored.

In short, though ASICs are logical candidates for the molecular acid taste sensor, there has been some conflicting evidence for their involvement in acid taste sensing, and reports so far have not employed ASIC KO strategies to investigate the in vivo role of ASICs in acid taste.

Hyperpolarization-Activated Cyclic Nucleotide-Gated Channels

HCN channels are members of the pore-loop cation channel superfamily [80], and in mammals, comprise four members, HCN1–4 that form tetrameric Na^+ and K^+ permeable channel complexes that are activated following membrane hyperpolarization, and which can be blocked by extracellular Cs^+ [5]. Each HCN is composed of a cytosolic N-terminal domain, a six TM core containing the gating mechanism and pore domain, and a large intracellular C-terminal domain that binds cyclic nucleotides [5].

Stevens et al. showed HCN1 and HCN4 to be expressed in gustducin-negative type III TRCs, observed a hyperpolarization-activated current that was enhanced by sour stimulation at the taste pore consistent with HCN kinetics, and proposed that HCN channels may act as sour taste receptors via gating by extracellular protons [63]. This contradicts the notion that intracellular acidification is the proximate stimulus, which would decrease the HCN activation threshold and slow channel opening [45, 81], and is verified by the observation that blocking HCNs in mice with Cs^+ did not affect acid-evoked responses [55]. It would thus appear unlikely that HCNs play a major role in acid taste.

K_2P Channels

K_2P channels are potassium leak channels that establish membrane potentials of all cells, and are especially important in determining the resting potential in neuronal cells. They are homodimeric proteins that each contain four TM and two pore-lining domains [34], and their conductances are sensitive to Ca^{2+}, cyclic nucleotides, or pH [66].

K_2P channels are expressed in the tongue tissue of mice [56] and rats [36], and in particular TWIK-1 and TASK-1 K_2P channels appear preferentially expressed in taste cells, and are modulated by changes in intracellular acidification [9, 34, 35, 41]. Pharmacological block of TASK-1 to simulate the effects of intracellular acidification appeared to enhance taste cell response to citric acid, consistent with a model in which intracellular acidification by acid tastants blocks the resting K^+ leak current of TASK-1 [56]. This evidence supports a role for channels in acid stimuli-mediated depolarization of type III taste cells, but it is still not known if K_2P channels are necessary or sufficient for sour sensing.

G-Protein-Coupled Receptors

GPCRs are ligand-activated, 7-TM proteins associated with heterotrimeric G proteins. GPCRs function in a wide variety of sensory mechanisms including sight and olfaction and signal through second messenger cascades. There are two known proton-sensing GPCRs, OGR1 and GPR4 [39]. GPR4 is a Gs-coupled GPCR and has been shown to function as a pH sensor in mice, affecting vascular and renal function when knocked out in mice [64, 75]. While it is not known if type III acid-sensing taste cells express acid-sensitive GPCRs or the associated signal transduction machinery required during activation, evidence for GPCR function in the crucial proton-sensing role in kidneys means that the function of GPCRs in acid taste is a possibility that should be more fully investigated.

Unknown Proton Channel or Transporter

Recently, Chang et al. demonstrated in PKD2L1-expressing sour taste cells from the mouse CvP the sufficiency of a Zn^{2+}-sensitive H^+ transduction current to generate action potentials independent of the presence of Na^+, and which is necessary for taste cell activation by strong acids. These findings point to a new class of proton channel or transporter which has not been previously considered that allows for direct proton entry into sour-sensing TRCs and is sufficient to depolarize these cells, and presumably elicit sour taste.

CO_2 Sensing by CA

One special case of acidic taste is the sensing of CO_2 by taste cells. In mammals, carbonation elicits both somatosensory and chemosensory responses in the oral cavity. CT nerve recordings showed responses to aqueous and gaseous CO_2 in a dose-dependent manner [6]. To identify TRCs sensing CO_2, taste nerve recording was performed on transgenic mice in which sweet or sour-sensing TRCs were specifically ablated by using cell-specific DTA expression. PKD2L1-DTA mice lacking sour-sensing type III cells did not show CT nerve responses to various concentration of CO_2, demonstrating that CO_2 is detected by the same TRCs as citric acid, acetic acid, and hydrochloric acid.

To identify a candidate receptor for carbonic acid, gene expression profiles in sour-sensing TRCs in wild-type mice were compared with those in taste buds of PKD2L1-DTA mice using a microarray technique. Car4, a member of CA family, was identified to be specifically expressed in sour-sensing TRCs. CAs are implicated in detecting CO_2 in various sensory systems including olfaction [20] by reversibly catalyzing the conversion of CO_2 to carbonic acid. Immunohistochemical staining using an anti-Car4 antibody revealed the expression of Car4 protein specifically in sour-sensing TRCs.

To verify the possibility that Car4 functions as a bona fide CO_2 sensor in taste system, CT nerve responses to CO_2 stimuli were examined using Car4 knockout mice and benzolamide, an inhibitor for CAs. In both cases, CT nerve response to CO_2, but not citric acid, was significantly reduced compared to wild-type mice. Thus, the detection of CO_2 is mediated by Car4 expressed in type III cells in mice, catalyzing the formation of carbonic acid, which then can be detected as a sour stimulus by the acid-sensing taste cells. Even though the Car4 pathway is a highly specialized mechanism for sensing of CO_2 as sour, it is a rare case where one specific gene product has been shown to be necessary for sensing a sour-tasting chemical, but it remains unknown if Car4 functions as a CO_2 sensor in humans.

Discussion

What Is Clear

Although the identity of the sour taste sensor is still not clear, there are some conclusions about this sensor that we can draw from the gathered evidence. Convergent experimental data indicate that the sour detection machinery is housed within Pkd2l1-expressing, presynaptic, type III TRCs or a subset of these cells. There is also evidence showing both dissociated protons and protonated acids stimulating the depolarization of these sour-sensing TRCs, which also supports the hypothesis that intracellular acidification is the proximate stimulus of sour taste.

Among the current list of candidate molecules, Pkd2l1 stands out as the only molecule so far that has been shown, by in vivo knockout experiments, to be involved in sour taste sensing, and could function, at least in the FuP papillae, as part of the sour-sensing mechanism in those cells.

What Is Not Clear/Needs Clarification

Many questions remain in the search for the sour sensor, including the identity of the sensor itself, but some of the other questions that remain may aid us in the search for the sour sensor. Chang et al. demonstrated the presence of a Zn^{2+}-sensitive, H^{+}selective channel that is sufficient to depolarize sour TRCs [7]. If those data are accurate, finding this proton channel could be key to understanding acid sensing in TRCs, both by elucidating the mechanism underlying strong acid taste and as a means for perturbation, since blocking or removing it may help determine the role of protonated acids in acid taste.

Another hole in our knowledge is the lack of evidence for the in vivo function of the other proposed sour-tasting candidates, as a number of them play important roles outside of their proposed role in sour taste. Since we have a reliable genetic marker for sour-sensing TRCs (Pkd2l1), sour TRC-specific knockout experiments will enable determination of their involvement in sour sensing. In addition, the intriguing "off-response" mediated by Pkd1l3/Pkd2l1 in TRCs in the CvP requires further investigation, as its presence implies a physiological function that we do not yet understand.

The Many Paths to Sour Sensing

There is also evidence that there could be multiple mechanisms involved in acid sensing in sour-sensing TRCs, foremost of which is the existence of two basic sour stimuli, protons and protonated acids, and further supported by the data showing that ASIC2 is involved in rat acid sensing, while not being expressed at all in sour-sensing TRCs in mouse, which may indicate that in there is not a universal or master acid sensor in these TRCs but two or more concurrent pathways that all contribute to acid sensing.

This seems to be the case in a human study, where Huque et al. examined FuP tissue in two patients suffering from sour ageusia compared to normal controls by performing RT-PCR search for likely candidate molecules for the sour sensor (ASIC/PKD/ENaC members) [25]. Both patients displayed abnormally high detection thresholds for citric acid (representing sour taste) but normal detection thresholds for the other taste modalities. While not a complete loss of detection, these patients were nevertheless categorized as having profound loss of sour taste. FuP cells from both patients were found to lack all ASICs as well as both Pkd1l3 and Pkd2l1, while ∂-ENaC appeared to be normally expressed, in contrast with normal sour tasters, whose FuP cells expressed ASICs and Pkd1l3/2l1. The simultaneous lack of ASICs and PKDs in both sour ageusic patients hints at a two-hit mechanism for their ageusia, and could mean that there are multiple paths to sour sensing.

Taken to the extreme, this could mean that the unique presynaptic nature of type III sour-sensing TRCs—their ability to be depolarized, to generate action potentials, and to release neurotransmitters—makes them the only cells that are able to transmit the broad-based cytosolic acidification of the lingual epithelium caused by sour tastants. In short, they are the only taste cells capable of transmitting this depolarizing acidification; and H^+, PKD, ASIC, HCN, and K_2P channels expressed in these cells work at the same time to depolarize the TRC, and basically only serve to enhance the sensitivity of the sour TRC to its intended stimuli.

On the other hand, the apparent existence of multiple pathways may simply be because all of the candidate molecules are simply accessory molecules, and the "master" acid sensor is simply yet to be identified. Since the most direct route currently proposed for TRC acidification is the proton channel, the discovery of such a channel and its genetic manipulation could allow us to determine the contribution of protons and protonated acids to sour TRC activation, and perhaps determine whether it is the "master" acid sensor.

References

1. Baeyens F, Vansteenwegen D, De Houwer J, Crombez G (1996) Observational conditioning of food valence in humans. Appetite 27(3):235–250. doi:10.1006/appe.1996.0049
2. Bartel DL, Sullivan SL, Lavoie EG, Sevigny J, Finger TE (2006) Nucleoside triphosphate diphosphohydrolase-2 is the ecto-ATPase of type I cells in taste buds. J Comp Neurol 497(1):1–12. doi:10.1002/cne.20954
3. Ben-Shahar Y (2011) Sensory functions for degenerin/epithelial sodium channels (DEG/ENaC). Adv Genet 76:1–26. doi:10.1016/B978–0-12–386481-9.00001–8
4. Bianchi L, Driscoll M (2002) Protons at the gate: DEG/ENaC ion channels help us feel and remember. Neuron 34(3):337–340
5. Biel M, Wahl-Schott C, Michalakis S, Zong X (2009) Hyperpolarization-activated cation channels: from genes to function. Physiol Rev 89(3):847–885. doi:10.1152/physrev.00029.2008
6. Chandrashekar J, Yarmolinsky D, von Buchholtz L, Oka Y, Sly W, Ryba NJ, Zuker CS (2009) The taste of carbonation. Science 326(5951):443–445. doi:10.1126/science.1174601
7. Chang RB, Waters H, Liman ER (2010) A proton current drives action potentials in genetically identified sour taste cells. Proc Natl Acad Sci U S A 107(51):22320–22325. doi:10.1073/pnas.1013664107
8. Chaudhari N, Roper SD (2010) The cell biology of taste. J Cell Biol 190(3):285–296. doi:10.1083/jcb.201003144
9. Chavez RA, Gray AT, Zhao BB, Kindler CH, Mazurek MJ, Mehta Y, Forsayeth JR, Yost CS (1999) TWIK-2, a new weak inward rectifying member of the tandem pore domain potassium channel family. J Biol Chem 274(12):7887–7892
10. Clapp TR, Yang R, Stoick CL, Kinnamon SC, Kinnamon JC (2004) Morphologic characterization of rat taste receptor cells that express components of the phospholipase C signaling pathway. J Comp Neurol 468(3):311–321. doi:10.1002/cne.10963
11. Damak S, Mosinger B, Margolskee RF (2008) Transsynaptic transport of wheat germ agglutinin expressed in a subset of type II taste cells of transgenic mice. BMC Neurosci 9:96. doi:10.1186/1471–2202-9–96

12. DeFazio RA, Dvoryanchikov G, Maruyama Y, Kim JW, Pereira E, Roper SD, Chaudhari N (2006) Separate populations of receptor cells and presynaptic cells in mouse taste buds. J Neurosci 26(15):3971–3980. doi:10.1523/JNEUROSCI.0515–06.2006
13. Finger TE, Danilova V, Barrows J, Bartel DL, Vigers AJ, Stone L, Hellekant G, Kinnamon SC (2005) ATP signaling is crucial for communication from taste buds to gustatory nerves. Science 310(5753):1495–1499. doi:10.1126/science.1118435
14. Ganzevles PG, Kroeze JH (1987a) Effects of adaptation and cross-adaptation to common ions on sourness intensity. Physiol Behav 40(5):641–646
15. Ganzevles PGJ, Kroeze JHA (1987b) The sour taste of acids. The hydrogen ion and the undissociated acid as sour agents. Chem Senses 12(4):563–576. doi:10.1093/chemse/12.4.563
16. Harvey RB (1920) The relation between the total acidity, the concentration of the hydrogen ion, and the taste of acid solutions. J Am Chem Soc 42(4):712–714. doi:10.1021/ja01449a005
17. Hayato R, Ohtubo Y, Yoshii K (2007) Functional expression of ionotropic purinergic receptors on mouse taste bud cells. J Physiol 584(Pt 2):473–488. doi:10.1113/jphysiol.2007.138370
18. Hirata K, Miyahara H, Kanaseki T (1988) Substance-P-containing fibers in the incisive papillae of the rat hard palate. Light- and electron-microscopic immunohistochemical study. Acta Anat (Basel) 132(3):197–204
19. Horio N, Yoshida R, Yasumatsu K, Yanagawa Y, Ishimaru Y, Matsunami H, Ninomiya Y (2011) Sour taste responses in mice lacking PKD channels. PLoS ONE 6(5):e20007. doi:10.1371/journal.pone.0020007
20. Hu J, Zhong C, Ding C, Chi Q, Walz A, Mombaerts P, Matsunami H, Luo M (2007) Detection of near-atmospheric concentrations of CO2 by an olfactory subsystem in the mouse. Science 317(5840):953–957. doi:10.1126/science.1144233
21. Huang YJ, Maruyama Y, Lu KS, Pereira E, Plonsky I, Baur JE, Wu D, Roper SD (2005) Mouse taste buds use serotonin as a neurotransmitter. J Neurosci 25(4):843–847. doi:10.1523/JNEUROSCI.4446–04.2005
22. Huang AL, Chen X, Hoon MA, Chandrashekar J, Guo W, Trankner D, Ryba NJ, Zuker CS (2006) The cells and logic for mammalian sour taste detection. Nature 442(7105):934–938. doi:10.1038/nature05084
23. Huang YA, Maruyama Y, Stimac R, Roper SD (2008) Presynaptic (Type III) cells in mouse taste buds sense sour (acid) taste. J Physiol 586(Pt 12):2903–2912. doi:10.1113/jphysiol.2008.151233
24. Huang YA, Dando R, Roper SD (2009) Autocrine and paracrine roles for ATP and serotonin in mouse taste buds. J Neurosci 29(44):13909–13918. doi:10.1523/JNEUROSCI.2351–09.2009
25. Huque T, Cowart BJ, Dankulich-Nagrudny L, Pribitkin EA, Bayley DL, Spielman AI, Feldman RS, Mackler SA, Brand JG (2009) Sour ageusia in two individuals implicates ion channels of the ASIC and PKD families in human sour taste perception at the anterior tongue. PLoS ONE 4(10):e7347. doi:10.1371/journal.pone.0007347
26. Inada H, Kawabata F, Ishimaru Y, Fushiki T, Matsunami H, Tominaga M (2008) Off-response property of an acid-activated cation channel complex PKD1L3-PKD2L1. EMBO Rep 9(7):690–697. doi:10.1038/embor.2008.89
27. Ishimaru Y (2009) Molecular mechanisms of taste transduction in vertebrates. Odontology 97(1):1–7. doi:10.1007/s10266–008-0095-y
28. Ishimaru Y, Matsunami H (2009) Transient receptor potential (TRP) channels and taste sensation. J Dent Res 88(3):212–218. doi:10.1177/0022034508330212
29. Ishimaru Y, Inada H, Kubota M, Zhuang H, Tominaga M, Matsunami H (2006) Transient receptor potential family members PKD1L3 and PKD2L1 form a candidate sour taste receptor. Proc Natl Acad Sci U S A 103(33):12569–12574. doi:10.1073/pnas.0602702103
30. Ishimaru Y, Katano Y, Yamamoto K, Akiba M, Misaka T, Roberts RW, Asakura T, Matsunami H, Abe K (2010) Interaction between PKD1L3 and PKD2L1 through their transmembrane domains is required for localization of PKD2L1 at taste pores in taste cells of circumvallate and foliate papillae. FASEB J 24(10):4058–4067. doi:10.1096/fj.10–162925
31. Johanningsmeiner SD, McFeeters RF, Drake M (2005) A hypothesis for the chemical basis for perception of sour taste. J Food Sci 70(2):R44–R48. doi:10.1111/j.1365–2621.2005.tb07111.x

32. Kawaguchi H, Yamanaka A, Uchida K, Shibasaki K, Sokabe T, Maruyama Y, Yanagawa Y, Murakami S, Tominaga M (2010) Activation of polycystic kidney disease-2-like 1 (PKD2L1)-PKD1L3 complex by acid in mouse taste cells. J Biol Chem 285(23):17277–17281. doi:10.1074/jbc.C110.132944
33. Kinnamon SC (2012) Taste receptor signalling—from tongues to lungs. Acta Physiol (Oxf) 204(2):158–168. doi:10.1111/j.1748-1716.2011.02308.x
34. Lesage F, Guillemare E, Fink M, Duprat F, Lazdunski M, Romey G, Barhanin J (1996) TWIK-1, a ubiquitous human weakly inward rectifying K+ channel with a novel structure. EMBO J 15(5):1004–1011
35. Lesage F, Lazdunski M (2000) Molecular and functional properties of two-pore-domain potassium channels. Am J Physiol Renal Physiol 279(5):F793–F801
36. Lin W, Burks CA, Hansen DR, Kinnamon SC, Gilbertson TA (2004) Taste receptor cells express pH-sensitive leak K+ channels. J Neurophysiol 92(5):2909–2919. doi:10.1152/jn.01198.2003
37. Lingueglia E (2007) Acid-sensing ion channels in sensory perception. J Biol Chem 282(24):17325–17329. doi:10.1074/jbc.R700011200
38. Lu Y, Ma X, Sabharwal R, Snitsarev V, Morgan D, Rahmouni K, Drummond HA, Whiteis CA, Costa V, Price M, Benson C, Welsh MJ, Chapleau MW, Abboud FM (2009) The ion channel ASIC2 is required for baroreceptor and autonomic control of the circulation. Neuron 64(6):885–897. doi:10.1016/j.neuron.2009.11.007
39. Ludwig MG, Vanek M, Guerini D, Gasser JA, Jones CE, Junker U, Hofstetter H, Wolf RM, Seuwen K (2003) Proton-sensing G-protein-coupled receptors. Nature 425(6953):93–98. doi:10.1038/nature01905
40. Lyall V, Alam RI, Phan DQ, Ereso GL, Phan TH, Malik SA, Montrose MH, Chu S, Heck GL, Feldman GM, DeSimone JA (2001) Decrease in rat taste receptor cell intracellular pH is the proximate stimulus in sour taste transduction. Am J Physiol Cell Physiol 281(3):C1005–C1013
41. Maingret F, Patel AJ, Lesage F, Lazdunski M, Honore E (1999) Mechano- or acid stimulation, two interactive modes of activation of the TREK-1 potassium channel. J Biol Chem 274(38):26691–26696
42. Makhlouf GBA (1972) Kinetics of the taste response to chemical stimulation: a theory of acid taste in man. Gastroenterology 63(1):67–75
43. Mantyh PW, Rogers SD, Honore P, Allen BJ, Ghilardi JR, Li J, Daughters RS, Lappi DA, Wiley RG, Simone DA (1997) Inhibition of hyperalgesia by ablation of lamina I spinal neurons expressing the substance P receptor. Science 278(5336):275–279
44. Medler KF, Margolskee RF, Kinnamon SC (2003) Electrophysiological characterization of voltage-gated currents in defined taste cell types of mice. J Neurosci 23(7):2608–2617
45. Munsch T, Pape HC (1999) Modulation of the hyperpolarization-activated cation current of rat thalamic relay neurones by intracellular pH. J Physiol 519(Pt 2):493–504
46. Nagy JI, Goedert M, Hunt SP, Bond A (1982) The nature of the substance P-containing nerve fibres in taste papillae of the rat tongue. Neuroscience 7(12):3137–3151
47. Nakamura K, Norgren R (1995) Sodium-deficient diet reduces gustatory activity in the nucleus of the solitary tract of behaving rats. Am J Physiol 269(3 Pt 2):R647–R661
48. Nelson TM, Lopezjimenez ND, Tessarollo L, Inoue M, Bachmanov AA, Sullivan SL (2010) Taste function in mice with a targeted mutation of the pkd1l3 gene. Chem Senses 35(7):565–577. doi:10.1093/chemse/bjq070
49. Ninomiya Y, Funakoshi M (1988) Amiloride inhibition of responses of rat single chorda tympani fibers to chemical and electrical tongue stimulations. Brain Res 451(1–2):319–325
50. Ohmoto M, Matsumoto I, Yasuoka A, Yoshihara Y, Abe K (2008) Genetic tracing of the gustatory and trigeminal neural pathways originating from T1R3-expressing taste receptor cells and solitary chemoreceptor cells. Mol Cell Neurosci 38(4):505–517. doi:10.1016/j.mcn.2008.04.011
51. Ohmoto M, Maeda N, Abe K, Yoshihara Y, Matsumoto I (2010) Genetic tracing of the neural pathway for bitter taste in t2r5-WGA transgenic mice. Biochem Biophys Res Commun 400(4):734–738. doi:10.1016/j.bbrc.2010.08.139

52. Oka Y, Butnaru M, von Buchholtz L, Ryba NJ, Zuker CS (2013) High salt recruits aversive taste pathways. Nature 494(7438):472–475. doi:10.1038/nature11905
53. Pan Z, Yang H, Reinach PS (2011) Transient receptor potential (TRP) gene superfamily encoding cation channels. Hum Genom 5(2):108–116
54. Price MP, McIlwrath SL, Xie J, Cheng C, Qiao J, Tarr DE, Sluka KA, Brennan TJ, Lewin GR, Welsh MJ (2001) The DRASIC cation channel contributes to the detection of cutaneous touch and acid stimuli in mice. Neuron 32(6):1071–1083
55. Richter TA, Caicedo A, Roper SD (2003) Sour taste stimuli evoke $Ca2^{+}$ and pH responses in mouse taste cells. J Physiol 547(Pt 2):475–483. doi:10.1113/jphysiol.2002.033811
56. Richter TA, Dvoryanchikov GA, Chaudhari N, Roper SD (2004a) Acid-sensitive two-pore domain potassium (K2P) channels in mouse taste buds. J Neurophysiol 92(3):1928–1936. doi:10.1152/jn.00273.2004
57. Richter TA, Dvoryanchikov GA, Roper SD, Chaudhari N (2004b) Acid-sensing ion channel-2 is not necessary for sour taste in mice. J Neurosci 24(16):4088–4091. doi:10.1523/JNEUROSCI.0653–04.2004
58. Roper SD (2007) Signal transduction and information processing in mammalian taste buds. Pflugers Arch 454(5):759–776. doi:10.1007/s00424–007-0247-x
59. Saper CB (2000) Hypothalamic connections with the cerebral cortex. Prog Brain Res 126:39–48. doi:10.1016/S0079–6123(00)26005–6
60. Scalera G, Spector AC, Norgren R (1995) Excitotoxic lesions of the parabrachial nuclei prevent conditioned taste aversions and sodium appetite in rats. Behav Neurosci 109(5):997–1008
61. Sowalsky RA, Noble AC (1998) Comparison of the effects of concentration, pH and anion species on astringency and sourness of organic acids. Chem Senses 23(3):343–349
62. Spector AC, Scalera G, Grill HJ, Norgren R (1995) Gustatory detection thresholds after parabrachial nuclei lesions in rats. Behav Neurosci 109(5):939–954
63. Stevens DR, Seifert R, Bufe B, Muller F, Kremmer E, Gauss R, Meyerhof W, Kaupp UB, Lindemann B (2001) Hyperpolarization-activated channels HCN1 and HCN4 mediate responses to sour stimuli. Nature 413(6856):631–635. doi:10.1038/35098087
64. Sun X, Yang LV, Tiegs BC, Arend LJ, McGraw DW, Penn RB, Petrovic S (2010) Deletion of the pH sensor GPR4 decreases renal acid excretion. J Am Soc Nephrol 21(10):1745–1755. doi:10.1681/ASN.2009050477
65. Sutherland SP, Benson CJ, Adelman JP, McCleskey EW (2001) Acid-sensing ion channel 3 matches the acid-gated current in cardiac ischemia-sensing neurons. Proc Natl Acad Sci U S A 98(2):711–716. doi:10.1073/pnas.011404498
66. Talley EM, Sirois JE, Lei Q, Bayliss DA (2003) Two-pore-Domain (KCNK) potassium channels: dynamic roles in neuronal function. Neuroscientist 9(1):46–56
67. Travers SP, Norgren R (1995) Organization of orosensory responses in the nucleus of the solitary tract of rat. J Neurophysiol 73(6):2144–2162
68. Ugawa S (2003) Identification of sour-taste receptor genes. Anat Sci Int 78(4):205–210. doi:10.1046/j.0022–7722.2003.00062.x
69. Ugawa S, Ueda T, Ishida Y, Nishigaki M, Shibata Y, Shimada S (2002) Amiloride-blockable acid-sensing ion channels are leading acid sensors expressed in human nociceptors. J Clin Invest 110(8):1185–1190. doi:10.1172/JCI15709
70. Ugawa S, Ueda T, Yamamura H, Nagao M, Shimada S (2005) Coexpression of vanilloid receptor subtype-1 and acid-sensing ion channel genes in the human trigeminal ganglion neurons. Chem Senses 30(Suppl 1):i195. doi:10.1093/chemse/bjh181
71. Waldmann R, Champigny G, Bassilana F, Heurteaux C, Lazdunski M (1997) A proton-gated cation channel involved in acid-sensing. Nature 386(6621):173–177. doi:10.1038/386173a0
72. Xie J, Price MP, Berger AL, Welsh MJ (2002) DRASIC contributes to pH-gated currents in large dorsal root ganglion sensory neurons by forming heteromultimeric channels. J Neurophysiol 87(6):2835–2843

73. Yamamoto K, Ishimaru Y, Ohmoto M, Matsumoto I, Asakura T, Abe K (2011) Genetic tracing of the gustatory neural pathway originating from Pkd113-expressing type III taste cells in circumvallate and foliate papillae. J Neurochem 119(3):497–506. doi:10.1111/j.1471–4159.2011.07443.x
74. Yamamoto T, Kawamura Y (1975) Dual innervation of the foliate papillae of the rat: an electrophysiological study. Chem Senses 1(3):241–244. doi:10.1093/chemse/1.3.241
75. Yang LV, Radu CG, Roy M, Lee S, McLaughlin J, Teitell MA, Iruela-Arispe ML, Witte ON (2007) Vascular abnormalities in mice deficient for the G protein-coupled receptor GPR4 that functions as a pH sensor. Mol Cell Biol 27(4):1334–1347. doi:10.1128/MCB.01909–06
76. Yang R, Montoya A, Bond A, Walton J, Kinnamon JC (2012) Immunocytochemical analysis of P2X2 in rat circumvallate taste buds. BMC Neurosci 13:51. doi:10.1186/1471–2202-13–51
77. Yarmolinsky DA, Zuker CS, Ryba NJ (2009) Common sense about taste: from mammals to insects. Cell 139(2):234–244. doi:10.1016/j.cell.2009.10.001
78. Yoshihara Y (2002) Visualizing selective neural pathways with WGA transgene: combination of neuroanatomy with gene technology. Neurosci Res 44(2):133–140
79. Yoshihara Y, Mizuno T, Nakahira M, Kawasaki M, Watanabe Y, Kagamiyama H, Jishage K, Ueda O, Suzuki H, Tabuchi K, Sawamoto K, Okano H, Noda T, Mori K (1999) A genetic approach to visualization of multisynaptic neural pathways using plant lectin transgene. Neuron 22(1):33–41
80. Yu FH, Yarov-Yarovoy V, Gutman GA, Catterall WA (2005) Overview of molecular relationships in the voltage-gated ion channel superfamily. Pharmacol Rev 57(4):387–395. doi:10.1124/pr.57.4.13
81. Zong X, Stieber J, Ludwig A, Hofmann F, Biel M (2001) A single histidine residue determines the pH sensitivity of the pacemaker channel HCN2. J Biol Chem 276(9):6313–6319. doi:10.1074/jbc.M010326200

Chapter 4
Function and Signaling of the pH-Sensing G Protein-Coupled Receptors in Physiology and Diseases

Lixue Dong, Zhigang Li and Li V. Yang

Introduction

The regulation of pH homeostasis in the body is tightly controlled by multiple physiological systems, such as through respiration, renal excretion, bone buffering, and metabolic modulation [1–4]. Acids are inevitably produced as end byproducts of cell metabolism. In aerobic metabolism, pyruvate is converted into carbonic dioxide and water along with the production of adenosine triphosphate (ATP) through oxidative phosphorylation. Based on the equilibrium, carbonic dioxide and carbonic acid are interconvertible. In anaerobic metabolism, pyruvate is directly lysed to lactic acid with the production of much less ATP. No matter whether the end product is carbonic dioxide or lactic acid, these weak acids must be transported out of the cells to maintain a stable intracellular pH homeostasis which is essential for numerous biochemical reactions. Cells have a large array of acid and base transporters, such as Na^+/H^+ exchangers, H^+-ATPase proton pump, and monocarboxylate transporters to maintain pH homeostasis [5–8]. After acids are transported out of the cells, they enter blood circulation and are removed from the body through respiratory exchange and renal excretion.

Although tightly regulated, pH homeostasis is disrupted in many pathological conditions. For example, systemic acidosis may occur in respiratory, renal, and metabolic diseases, sepsis shock, and critically ill patients [1–4, 9]. Local acidosis frequently exists in ischemic tissues, solid tumors, inflammatory tissues such as arthritis and asthma, and other conditions [10–16]. Because of inadequate blood perfusion and hypoxia, cells have to switch to glycolytic metabolism. In the case

L. V. Yang (✉)
Department of Oncology, Department of Internal Medicine,
Department of Anatomy and Cell Biology, Brody School of Medicine,
East Carolina University, Greenville, NC 27834, USA
e-mail: yangl@ecu.edu

L. Dong · Z. Li
Department of Oncology, Brody School of Medicine,
East Carolina University, Greenville, NC 27834, USA

J-T. A. Chi (ed.), *Molecular Genetics of Dysregulated pH Homeostasis*,
DOI 10.1007/978-1-4939-1683-2_4

of cancer, tumor cells intrinsically have increased glycolysis even in the presence of oxygen, a phenomenon called "Warburg effect" [17]. Glycolytic metabolism of cells produces excessive amount of lactic acid, which is transported out of cells, together with other proton sources such as from carbonic acid, and ATP hydrolysis [18, 19], causing acidosis in local tissues. Studies show that the interstitial pH can fall to as low as 5.5–7.0 in ischemic tissues, solid tumors, and inflammatory tissues [10–12, 14, 15]. Acidosis has profound effects on pathophysiology and can regulate cell death, proliferation, blood vessel function, immunity, inflammation, cancer progression and therapeutic response, pain sensation, and others [10, 11, 14–16, 20, 21].

Acidic extracellular pH can be sensed by several types of cell-surface channels and receptors [22]. Among these, acid-sensing ion channels (ASICs) and transient receptor potential (TRP) ion channels have been most extensively studied. Upon activation by acidic extracellular pH, the acid-sensitive channels become permeable to ions and regulate cell activity. In addition to proton-activated ion currents, early studies by Smith et al. showed that the acidic extracellular pH can induce inositol polyphosphate formation and calcium efflux in cells, similar to the effects of bradykinin [23]. In contrast, acidification of intracellular pH cannot stimulate calcium efflux. The authors suggested that the acidic pH may cause the protonation of a critical functional group, possibly the imidazole of histidine, of a cell-surface receptor [23]. More studies from the same group further substantiated that the acidic extracellular pH evokes calcium mobilization through a putative cell-surface receptor [24], and many other groups observed that the acidic extracellular pH can induce calcium efflux in various cell types [25–28]. However, the molecular identity of the putative acid-sensing receptor was not known. The seminal work by Ludwig et al. discovered that a family of GPCRs can be activated by acidic extracellular pH [29]. When Human ovarian cancer G protein-coupled receptor 1 (OGR1) (GPR68), one of the proton-sensing GPCRs, is activated by acidic extracellular pH, inositol phosphate formation and calcium efflux through the G_q-coupled pathway are detected in cells. Furthermore, the authors demonstrated that several histidine residues are important for the proton sensing function of the OGR1 receptor [29]. The proton-sensing GPCRs identified by Ludwig and others possess all the biochemical features initially observed by Smith et al., such as acidic extracellular pH-induced activation, inositol phosphate formation, and calcium efflux [23]. In addition to OGR1 (GPR68), three other GPCRs, including GPR4, TDAG8 (GPR65), and G2A (GPR132), have also been identified as acid sensors [20, 29–40]. These receptors represent a unique family of GPCRs which are responsive to extracellular pH change.

The proton-sensing GPCRs are activated by acidic extracellular pH and transduce downstream signals through the G_s, G_q, and G_{13} pathways. A number of research groups, including our own, have shown that the acid-sensing GPCRs play roles in cardiovascular, immune, renal, nervous, skeletal, and respiratory systems, and inflammation and cancer biology [20, 21, 29–44]. Historically, this family of receptors, including GPR4, OGR1, TDAG8, and G2A, has been proposed as receptors for bioactive lipids [45–48]. However, several of the initial papers have been withdrawn because the ligand binding and other key data could not be reproduced

[46–48]. In the case of G2A, some lysophosphatidylcholine (LPC)-induced effects have been reproducibly demonstrated [49–59], but it is unclear whether LPC acts on G2A through direct or indirect mechanisms. On the other hand, the pH-sensing function of this family of receptors has been confirmed by a large number of studies. Here, we will focus on the pH-sensing function of the receptors, and discuss the biochemical signaling and biological function of these receptors in physiology and diseases.

GPR4

G protein-coupled receptor 4 (GPR4) was originally identified as an orphan GPCR. The human GPR4 gene is localized in chromosome 19q13.3. There are two human GPR4 mRNA isoforms which are expressed in many tissues, with the highest expression level in the lungs and lower levels in the kidneys, heart, skeletal muscle, liver, and pancreas [60]. Lum et al. showed that GPR4 is expressed in two immortalized endothelial cell lines, with low mRNA expression in human brain microvascular endothelial cells (HBMEC) and high expression in human dermal microvascular endothelial cells (HMEC). Also, inflammatory stimuli, including tumor necrosis factor-α (TNFα) and hydrogen peroxide can induce GPR4 expression in HBMEC, but not in HMEC [61]. Moreover, several research groups, including our own, showed that GPR4 has high levels of expression in a variety of primary endothelial cells, including human umbilical vein endothelial cells (HUVEC), human pulmonary artery endothelial cells (HPAEC), human aortic endothelial cells (HAEC), and human lung microvascular endothelial cells (HMVEC-L), whereas the other three members of this family (OGR1, TDAG8, and G2A) have barely detectable expression [40, 62, 63]. It has been shown that GPR4 expression in HUVEC is reduced by the treatment of acidic pH [20]. In tumors, Sin et al. showed that GPR4 is overexpressed in certain percentage of various human cancer tissues, including breast, ovarian, colon, liver, and kidney tumors [64]. On the other hand, Wyder et al. showed that there is almost no GPR4 expression in several human and mouse tumor cell lines [63]. We also examined GPR4 expression in more than a dozen human cancer cell lines and found its expression level to be low in these cancer cells (Dong L. et al. unpublished data). As tumor tissues consist of cancer cells, endothelial cells, and many other types of stromal cells, it is important to define which cell types express GPR4 in the tumors.

GPR4 was initially reported as a receptor for bioactive lipids sphingosylphosphorylcholine (SPC) and LPC [48]. However, the original paper was retracted after failing to reproduce the receptor binding data. Ludwig and colleagues instead observed that GPR4 responds to extracellular pH change and induces cyclic adenosine 5′-monophosphate (cAMP) formation through Gs proteins, indicating that GPR4 works as a receptor of protons [29]. The results regarding the relationship between the lipid molecules and GPR4 are controversial. On one hand, Bektas et al. suggested that SPC and LPC are not the ligands for GPR4 [65]; The authors showed that in GPR4 expressing cells, SPC, LPC, and other related lysophospholipids cannot induce GPR4 internalization from cell membrane or induce β-arrestin translocation from cytosol to plasma membrane. In addition, these lysolipids are not able to stimulate GTPγS binding to membranes or activate ERK1/2. On the other hand, several studies showed that GPR4 is involved in a variety of cellular activities that are induced by SPC and LPC, such as SPC-induced endothelial tube formation [62], LPC-mediated endothelial barrier dysfunction [66], LPC-stimulated monocyte transmigration through endothelial cell monolayer [67], and LPC-induced expression of adhesion molecules in rat endothelial cells [68]. However, since the receptor–ligand relationship between GPR4 and SPC or LPC cannot always be confirmed, these lipid molecules might only function through GPR4 indirectly. Meanwhile, ligand-independent activation of GPR4 was also reported. Bektas et al. showed that overexpression of GPR4 inhibits SPC, sphingosine-1-phosphate (S1P), or EGF-induced ERK1/2 activation, in the absence of added ligands [65]. Moreover, Sin et al. showed that overexpression of GPR4 in HEK293 cells results in transcriptional activation of multiple signaling pathways without the addition of exogenous ligand [64]. More recent studies, however, suggested that extracellular protons at least partly work as ligands for those previously proposed ligand-independent responses in GPR4-expressing cells. Intriguingly, the studies demonstrated that upon acidic extracellular pH activation, GPR4 couples to G_s/cAMP, $G_{12/13}$/Rho, and to a lesser extent, $G_{q/11}$/phospholipase C (PLC) signaling pathways [44]. It is also shown that protonation of each one of histidine residues at 79, 165, and 269 in GPR4 is critical for GPR4 activation and coupling to multiple intracellular signaling pathways [42]. To date, it is widely recognized that GPR4 is a proton-sensing GPCR involved in a variety of biological processes.

Acidic tissue microenvironment commonly exists in many pathophysiological conditions [10–16]. Endothelial cells in blood vessels under these conditions are frequently exposed to an acidic extracellular pH. GPR4 is expressed at higher level in endothelial cells than the other three family members; therefore, it may play pivotal roles in mediating the effects of acidosis on endothelial cells. One study reported that GPR4 is important for tube formation, cell survival, and proliferation in HUVEC and immortalized human microvascular endothelial cells (HMEC-1), albeit the effect of pH changes on these processes was not examined [62]. However, in this study, the authors could not observe an increase of cAMP production upon acidic pH treatment in parental or GPR4-overexpressing HMEC-1 cells. Therefore, they argued that GPR4 is not a proton sensor in these endothelial cells. In contrast, Chen et al. showed that extracellular acidosis, either isocapnic or hypercapnic, can induce cAMP production in HUVEC, and this induction effect is much stronger in GPR4-overexpressing HUVEC, suggesting a GPR4-dependent proton-sensing function [20]. In line with these results, other studies also demonstrated an increase of cAMP production in HUVEC in response to extracellular acidosis [40, 63]. Furthermore, extracellular acidosis, either isocapnic or hypercapnic, can activate GPR4 to increase the adhesion of HUVEC to leukocytes mainly through the G_s/cAMP/Epac pathway [20].

The important role of GPR4 in blood vessels is further corroborated by the phenotype of GPR4-null mice. There are two reports in which GPR4-deficient mice were generated, based on different mouse strains. Phenotype of GPR4-null mice was initially reported in a C57BL/6 and 129 mixed genetic background [40]. The authors observed that GPR4-null adult mice are viable and fertile, and show grossly normal phenotype. However, the litter size is slightly smaller and the perinatal mortality rate is higher in GPR4$^{-/-}$ mice, which is probably caused by respiratory distress from lung epithelial metaplasia. Further examination revealed a partial penetrance of various degrees of spontaneous hemorrhaging in a small fraction of GPR4 knockout embryos and neonates, likely due to defects in blood vessels. Indeed, histological analysis showed that in the hemorrhagic GPR4-null mice, there are dilated, tortuous, and poorly organized blood vessels with decreased smooth muscle cell coverage, particularly, in small blood vessels. In addition, mesangial cell coverage of kidney glomeruli is also reduced in GPR4-null neonates [40]. The other study on GPR4-deficient mice reported similar adult mouse phenotype with no gross abnormalities, but did not observe the perinatal phenotype in GPR4-null mice. These discrepancies might be due to difference in animal strain backgrounds, breeding and backcrossing conditions, and gene knockout constructs [63]. In agreement with the critical role of GPR4 in blood vessels, the response to vascular endothelial growth factor (VEGF)- but not basic fibroblast growth factor (bFGF)-driven angiogenesis is significantly reduced in GPR4-null mice. Moreover, blood vessels in syngeneic tumors grown in GPR4-null mice in BABL/c genetic background, compared with that in wild-type mice, are more fragmented and fragile, further suggesting a role of GPR4 in the regulation of blood vessel stability [63].

Kidney is an organ essential for the maintenance of pH homeostasis by tightly controlling acid and base excretion. Therefore, it is of great importance for kidney to sense and respond to systemic pH change [38]. With relatively abundant expression in kidney, the role of GPR4 in renal cells has been investigated. Sun et al. examined GPR4 expression in mouse kidney and found that GPR4 is expressed at high levels in the kidney cortex and in the outer and inner medulla [38]. With regard to nephron segments, GPR4 is expressed in isolated kidney collecting ducts, and in cultured mouse outer and inner medullary collecting duct cells (mOMCD1 and mIMCD3). Similar as in endothelial cells, acidic pH also induces an accumulation of intracellular cAMP in mOMCD1 cells through the activation of GPR4 [38]. The role of GPR4 as a pH sensor in kidney was further validated in GPR4-deficient mice, which exhibited decreased kidney net acid secretion, spontaneous metabolic acidosis, and defective response to acid challenge, compared with wild-type mice [38]. The α-subunit of H^+-K^+-ATPase (HKα_2) plays important roles in maintaining systemic acid–base homeostasis and defending against metabolic acidosis. Codina et al. showed that chronic acidosis increases protein kinase A (PKA) activity and the protein expression of HKα_2 without changing HKα_2mRNA abundance in kidney cells. Ectopic overexpression of GPR4 further increases both basal and acidosis-stimulated PKA activity and HK$\alpha_2$expression, indicating the activation of GPR4 by acidosis in kidney cells [69].

Tumor microenvironment is characterized by hypoxia and acidosis due to disorganized tumor vasculature and increased glycolysis of cancer cells (Warburg effect) [10]. The proton-sensing GPCRs may be involved in the regulation of cancer progression as they are activated by acidic pH in the tumor microenvironment. In one study, it is reported that overexpressed GPR4 induces oncogenic transformation of immortalized NIH3T3 fibroblasts, suggesting that GPR4 has tumor promoting activity in NIH3T3 cells. However, the effects of pH were not examined in this study [64]. In contrast, Castellone et al. demonstrated that the activation of GPR4 by acidic pH significantly inhibits tumor cell migration and invasion in vitro [30]. Furthermore, GPR4 overexpression substantially suppresses the pulmonary metastasis of malignant B16F10 mouse melanoma cells in vivo. Therefore, GPR4 might function as a tumor metastasis suppressor [30]. Utilizing 3D morphology analysis, Zhang et al. revealed that B16F10 mouse melanoma cells with GPR4 overexpression have reduced membrane protrusions and increased mitochondrial surface area, which is consistent with decreased cell migration ability and increased maximal capacity of mitochondrial oxygen consumption rate, respectively [70].

OGR1 (GPR68)

OGR1, also referred as GPR68, was initially identified in an ovarian cancer cell line, HEY. OGR1 is widely expressed in many tissues, including spleen, testis, small intestine, peripheral blood leukocytes, heart, placenta, brain, lung, bone, and kidney. Its expression is absent or very low in other tissues, such as thymus, prostate, ovary, colon, liver, pancreas, or skeletal muscle [71]. Within the proton-sensing GPCR family, OGR1 shares the highest homology (49–54 %) with GPR4.

Similar to GPR4, OGR1 was first proposed to be a high-affinity receptor for the bioactive lipid SPC [47]. The binding of SPC to OGR1 was reported to result in transient intracellular calcium increase, activation of p42/44 mitogen-activated protein kinases (MAPK), and inhibition of cell proliferation. However, the original article was later retracted, concerning about a portion of the published data possibly being falsified [47]. Several other groups also failed to observe the agonistic activity of SPC for OGR1 in various cells overexpressing OGR1 [29, 72]. On the other hand, an antagonistic action of high concentrations of SPC was reported in acidic pH-induced OGR1 activation [72]. It was also suggested that the antagonistic regulation by micromolar concentrations of lipids might be due to nonspecific effects [37].

More recently, Ludwig et al. demonstrated that OGR1 functions as a proton-sensing GPCR [29]. OGR1 is inactive at alkaline pH (pH 7.8), whereas it is fully activated at slightly acidic pH (pH 6.4–6.8). The activation of OGR1 by protons stimulates inositol phosphate formation, intracellular calcium efflux, and the SRE (serum response element) reporter gene expression, indicating the coupling of OGR1 to G_q and G_{13} proteins. The study also identified OGR1 expression in bone cells, including human MG63 osteosarcoma cells, rat osteoblasts, and rat osteocytes

[29]. The discovery of OGR1 as a pH-sensing receptor triggered a considerable interest of research in bone biology, as will be discussed below. The activation of OGR1 by acidic extracellular pH can also stimulate cAMP production, reflecting the activation of G_s protein and adenylyl cyclase [72, 73]. However, the underlying signaling pathways responsible for the cAMP production are different based on two studies from Okajima's group. In one study, the induced cAMP production was reported to result from the activation of adenylyl cyclase through PLC/ERK/phospholipase A_2/COX/PGI_2 pathway [73], whereas in the other study, cAMP production is the direct result of OGR1 coupling to G_s proteins [72].

Bone plays an important role in regulating acid–base balance and maintaining pH homeostasis. Chronic metabolic acidosis increases urine Ca^{2+} excretion and results in the depletion of bone calcium store [4]. Excessive calcium loss can cause osteopenia such as osteoporosis. Therefore, systemic acidosis has detrimental effects on bone. A number of studies demonstrated that OGR1 is involved in various aspects of extracellular acidosis-induced bone loss. NFATc1 is a transcription factor essential for osteogenesis and is highly induced by RANKL (receptor activator of NF-kappa B ligand) signaling during osteoclast differentiation [74]. It is reported that, similar to RANKL, acidic extracellular pH significantly induces NFATc1 activation in rat and rabbit osteoclasts through PLC/Ca^{2+}/calcineurin pathway [75]. Although its specific physiological role was not examined, OGR1 was suggested to mediate acidosis-stimulated NFATc1 activation and osteoclastic resorption [75]. When CSF-1 is injected to restore osteoclastogenesis in the CSF-1-null *toothless* ($csf1^{tl}/csf1^{tl}$) osteopetrotic rat, OGR1 is upregulated by greater than sixfold [76]. OGR1 is also strongly upregulated in mouse bone marrow mononuclear cells and pre-osteoclast-like cells treated with RANKL to induce osteoclast differentiation. Moreover, inhibition of OGR1 by antibody or small interfering RNA (siRNA) abolishes RANKL-induced osteoclastogenesis [76]. OGR1 is also involved in extracellular acidosis-induced Ca^{2+} signaling and osteoclast survival in an NFAT-independent, protein kinase C (PKC)-dependent manner [77]. Consistent with in vitro studies, fewer osteoclasts derived from bone marrow cells were detected in OGR1-deficient mice. A weak pH-dependent osteoclast survival effect was also observed. However, overall bone structure is not affected in OGR1-null mice [78]. Other studies showed that OGR1 is responsible for acidosis-induced COX-2 expression and prostaglandin E_2 production in a human osteoblastic cell line (NHOst) through the $G_{q/11}$/PLC/PKC pathway [79]. Frick et al. showed that both metabolic and respiratory acidosis can induce a transient increase in $[Ca^{2+}]_i$ in primary bone cells [31, 80]. Inhibition of intracellular calcium release by IP_3 (Inositol 1,4,5-triphosphate) signaling inhibitors abolishes acidosis-induced COX-2 and RANKL expression and bone resorption [81].

In the vascular system, acidosis induces a variety of responses such as blood vessel dilation and relaxation, cellular cAMP accumulation, $[Ca^{2+}]_i$ alteration, and inhibition of vascular smooth muscle cell proliferation and migration [5, 82–85]. However, the underlying mechanisms were unclear. Tomura et al. showed that an acute acidic extracellular pH induces inositol phosphate production, $[Ca^{2+}]_i$ elevation, cAMP accumulation, and prostaglandin I_2 (PGI_2) production in human aortic smooth muscle cells (AoSMCs). The authors also showed that OGR1 is the major receptor involved in these effects as knockdown of OGR1 expression inhibits these events [73]. Subsequently, the same group found that chronic extracellular acidification exerts multiple effects on the functions of AoSMCs. Acidosis-induced COX-2 expression, PGI_2 production, and MKP-1 expression is mediated by OGR1, whereas the other effects, such as increased PAI-1 expression and inhibition of AoSMC proliferation, are OGR1-independent [86].

Acidic tissue microenvironment also exists in inflammatory airway diseases, in which the pH drops to as low as 5.2–7.1, compared with pH 7.5–7.7 in the airway of healthy subjects [12, 87–89]. Airway acidification is associated with the pathophysiology of inflammatory airway diseases, such as asthma [12, 87, 89]. Ichimonji and colleagues showed that OGR1/$G_{q/11}$plays a role in asthmatic responses by mediating extracellular acidification-induced production of the proinflammatory cytokine IL6 and intracellular Ca^{2+} efflux in human airway smooth muscle cells (ASMCs) [90]. The same group also found that the extracellular acidification alone or with transforming growth factor (TGF)-β stimulates connective tissue growth factor (CTGF) production through the OGR1/$G_{q/11}$/IP_3/Ca^{2+} pathway in human ASMCs [91]. Moreover, Saxena and colleagues reported that human ASMCs respond to small reductions in extracellular pH, leading to the activation of multiple signaling pathways as well as Ca^{2+}-dependent cell contraction, which are mainly mediated through OGR1 [43]. Excess secretion of mucus is common in inflammatory airway diseases and is exacerbated by acidic stress [87, 89]. Liu et al. showed that acid-induced calcium mobilization and mucin5AC hypersecretion is mediated through the OGR1/$G_{q/11}$/PLC pathway in human airway epithelium [92]. Therefore, OGR1 may play important roles in the pathophysiology of inflammatory airway diseases and represent a potential therapeutic target in lung diseases.

The roles of OGR1 in tumor biology have been reported. OGR1, together with G2A, is expressed in a human medulloblastoma cell line (DAYO) as well as in medulloblastoma tissues from patients. Although a direct involvement of these two receptors was not shown, Huang et al. reported that extracellular acidification evokes spatially and temporally distinct Ca^{2+} signals, which, in turn, activates the MEK/ERK pathway as a possible mechanism by which acidosis affects cell proliferation [93]. Singh et al. showed that OGR1 inhibits tumor metastasis when the receptor is overexpressed in prostate cancer cells, but has no effect on primary tumor growth, suggesting that OGR1 might function as a metastasis suppressor [94]. In addition, overexpression of OGR1 in human ovarian cancer cells inhibits cell proliferation and migration and increases cell adhesion to extracellular matrix, also indicating a role of OGR1 as a tumor suppressor [95]. On the other hand, Li et al. showed that melanoma formation is significantly inhibited in OGR1-deficient mice, suggesting a tumor-promoting function of OGR1 [78]. In a more recent report, they demonstrated that OGR1 deficiency also reduces the tumor formation of prostate cancer cells in mice. The underlying mechanisms were attributed to a requirement of OGR1 in myeloid-derived cells to mediate tumor cell-induced immunosuppression [96].

OGR1 is also involved in a number of other biological processes. Recently, Mohebbi and colleagues reported that acidic pH activation of OGR1 stimulates the activity of two major proton transport systems, Na^+/H^+ exchanger 3 (NHE3) and H^+-ATPase, in renal epithelial HEK293 cells. Deletion of OGR1 also affects NHE3 activity upon acidic extracellular pH treatment in proximal tubules isolated from kidneys of OGR1-deficient mice, suggesting a role of OGR1 in the regulation of renal acid transport [97]. Moreover, Nakakura et al. reported that high glucose-induced insulin secretion is decreased in OGR1-deficeint mice. Concordantly, in vitro studies demonstrated that the OGR1/$G_{q/11}$pathway activated by acidic pH augments insulin secretion in response to high concentration of glucose [98].

TDAG8 (GPR65)

The T cell death-associated gene 8, TDAG8 (also known as GPR65), was initially cloned as an orphan GPCR, which is upregulated during thymocyte apoptosis [99, 100]. Upon T cell activation by anti-T cell receptor antibody or by phorbol 12-myristate 13-acetate and ionomycin, the expression of TDAG8 is substantially induced [99, 100]. Also, TDAG8 is greatly induced upon glucocorticoid-induced apoptosis, which was demonstrated not only in WEHI7.2 and S49.A2 murine T cell lymphoma cell lines but also in primary thymocytes [99, 101–104]. In human, TDAG8 gene is mainly expressed in lymphoid tissues, including spleen, lymph nodes, leukocytes, and thymus; similarly, mouse has high expression of TDAG8 in thymus and spleen [99, 100, 103]. Together, the expression pattern of TDAG8 suggests that it may play a role in immune regulation.

Initially, psychosine was identified as a ligand of TDAG8 as this glycosphingolipid inhibits forskolin-evoked cAMP accumulation and induces calcium mobilization in TDAG8-expressing cells [45]. Malone et al. showed that psychosine significantly enhances dexamethasone-induced apoptosis via TDAG8 [102]. More recent studies demonstrated that TDAG8 is a proton-sensing GPCR. As a downstream signaling molecule of TDAG8, cAMP is significantly increased by acidic extracellular pH in several cell types transfected with TDAG8, but markedly attenuated by copper ions (inhibitor of proton-sensing GPCRs) and in cells expressing mutated TDAG8 [33, 36, 39, 101]. Phosphorylation of CREB downstream of cAMP is also induced by acidic pH treatment in TDAG8 overexpressing Chinese hamster ovary (CHO) cells [105]. In cell-free system, both GTPγS binding activity and adenylyl cyclase activity are greatly induced by acidic extracellular pH in TDAG8 overexpressing cell membrane fractions but not in control cell membrane fractions [39], further demonstrating the pH-sensing activity of TDAG8. Ishii et al. showed that acidic extracellular pH induces TDAG8-dependent stress fiber formation and Rho activation in CHO-S cells, suggesting the coupling of TDAG8 to $G_{12/13}$ proteins [33]. GPCR internalization is a common phenomenon upon ligand-induced receptor activation. When cells were treated with acidic pH, the internalization of TDAG8 was detected, suggesting protons as a ligand of TDAG8 [33]. In contrast to the previously reported agonist effects [45], Wang et al. identified psychosine as a TDAG8 antagonist as it inhibits cAMP accumulation triggered by extracellular protons, an agonist of TDAG8 [39].

Several studies have demonstrated the function of TDAG8 in inflammation and immunity. TDAG8 responds to extracellular acidification and inhibits the production of proinflammatory cytokines, such as TNFα and IL-6, in mouse macrophages through the G_s protein/cAMP/PKA signaling pathway [106]. Also, TDAG8 was suggested to be partly involved in the glucocorticoid-induced antiinflammatory actions, as TDAG8 deficiency attenuates dexamethasone-induced inhibition of inflammatory cytokine production (TNFα) in mouse peritoneal macrophages at acidic pH environment [101]. Other research showed that TDAG8 is highly expressed in eosinophils (hallmark cells of asthmatic inflammation) and is induced in asthmatic inflammation [34]. In murine asthma models, TDAG8 was shown to increase the viability of eosinophils at acidic pH through the cAMP pathway [34]. In addition to asthma, TDAG8 may also play a role in arthritis. TDAG8-null mice, compared with wild-type mice, exhibit more severe arthritis induced by anti-type II collagen antibody and a mild exacerbation of collagen-induced arthritis [107]. It was recently reported that a small molecule, BTB09089, functions as a TDAG agonist and simulates cAMP production in a TDAG8-dependent manner in mouse splenocytes [108]. This compound also inhibits inflammatory cytokine production in mouse T cells and macrophages [108]. In addition, TDAG8 is also involved in acidosis-associated inflammatory pain. TDAG8 is induced by complete Freund's adjuvant (CFA) injection in the neurons of pain-relevant loci (dorsal root ganglia) and sensitizes the transient receptor potential vanilloid 1 (TRPV1) response to capsaicin [109]. Although TDAG8 is involved in various inflammatory responses, a study showed the dispensability of TDAG8 in immune development [103]. TDAG8 knockout mice have normal immune development and intact major immune functions [103]. Furthermore, T lymphocytes explanted from TDAG8 knockout mice show normal immune phenotypes and psychosine-induced inhibition of cytokinesis. Compared with wild-type mouse thymocytes, TDAG8-deficient thymocytes exhibit similar glucocorticoid-induced apoptosis [103].

In addition to inflammation, TDAG8 is also reported to play a role in tumor biology. Human TDAG8 gene is located in the chromosome 14q31–32.1, a region associated with T cell lymphoma or leukemia [100]. Overexpression of TDAG8 was shown to confer transformed phenotypes, including refractile cell shape, foci formation, and low serum resistance, in NIH3T3 immortalized fibroblasts [64]. Overexpression of TDAG8 in the NMuMG mammary epithelial cell line induces tumor growth in nude mice [64]. Ihara et al. also reported tumor-promoting function of TDAG8 [32]. Overexpression of TDAG8 in mouse Lewis lung carcinoma (LLC) cells enhances tumor cell resistance to acidic condition, while knockdown of TDAG8 in NCI-H460 human non-small cell lung cancer cells attenuates cell survival in acidic environment [32]. In vivo study showed that mice exhibit larger tumors when injected with TDAG8-overexpressing LLC cancer cells compared with the mice injected with cancer cells expressing endogenous TDAG8, while the knockdown of TDAG8 in NCI-H460 lung cancer cells inhibits tumor development in mouse model [32]. Recently, TDAG8 has been demonstrated to sense the acidic tumor microenvironment and promote evasion of apoptosis through activating the MEK/ERK pathway under glutamine starvation in WEHI7.2 and CEM-C7 T cell lymphoma cell lines [110]. In contrast to the tumor-promoting effects described above, tumor-suppressing activities of TDAG8 have also been reported. TDAG8, as a proapoptotic gene, promotes glucocorticoid-induced apoptosis in murine lymphoma cells [102]. These observations suggest that the function of TDAG8 in tumor biology may be cell type and context dependent.

G2A (GPR132)

G2A (G2 accumulation) was originally identified as a DNA damage and stress-induced GPCR, with high expression level in immature T and B lymphocyte progenitors [111]. In mouse bone marrow B cells, G2A is highly expressed in pro-B cells with a low level in pre-B and immature B cells. In thymocytes, G2A is expressed in all stages of intrathymic maturation [111]. G2A is induced by various DNA-damaging agents, such as hydroxyurea, 5′-fluorouracil, cytosine arabinoside, etoposide, taxol, doxorubicin, x-ray, and UV (ultraviolet). Overexpression of G2A in NIH3T3 fibroblasts results in cell cycle blockade in the G2/M phase [111].

Initial studies showed that G2A functions as a tumor suppressor. BCR-ABL, a chimeric oncogene generated by chromosomal translocation, is critical for the pathogenesis of chronic myelogenous leukemia and acute lymphocytic leukemia. G2A is a BCR-ABL target gene that is transcriptionally induced in mouse bone marrow cells transduced with the BCR-ABL oncogene [111]. Interestingly, overexpression of G2A attenuates the transformation potential of BCR-ABL and causes cell cycle arrest in the G2/M phase, suggesting a tumor-suppressing function of G2A [111]. Furthermore, when G2A-null mouse bone marrow cells were transduced with the BCR-ABL gene and transplanted into irradiated recipient mice, higher efficiency of leukemogenesis was observed in comparison to the recipients of BCR-ABL transduced wild-type bone marrow [112]. Thus, G2A is proposed to be a tumor suppressor, which triggers growth inhibitory signals in response to malignant transformation by certain oncogenes. However, other studies suggest that G2A may be oncogenic. Cotransfection of G2A dramatically increases the transformation efficiency by a weakly oncogenic form of Raf-1in NIH3T3 fibroblasts [113]. Even overexpression of G2A alone, coupled to G_{13} and Rho GTPase, induces various phenotypes characteristic of oncogenic transformation in NIH3T3 fibroblasts, including suppression of contact inhibition, anchorage-independent growth, reduced dependence on serum, and tumor formation in mice [113]. Therefore, the function of G2A in tumorigenesis is complex and seems to be cell-type dependent.

LPC was initially identified as a high-affinity ligand for G2A [46], but the paper describing this data was retracted by the authors because the specific binding of LPC to G2A-expressing cells could not be reproduced. Other studies suggested that instead of directly binding to G2A, LPC regulated the intracellular sequestration and cell surface redistribution of G2A [58]. More recently, G2A is proposed to be a proton-sensing GPCR. Acidic extracellular pH increases inositol phosphate accumulation in G2A-expressing cells [35]. Nevertheless, another study showed that G2A is less sensitive to acidic pH as measured by inositol phosphate and cAMP accumulation compared with other members of proton-sensing GPCRs (OGR1, GPR4, and TDAG8) [36]. Moerover, G2A is dispensable for the production of cAMP by acidic pH treatment in mouse thymocytes and splenocytes, while another proton-sensing GPCR TDAG8 is indispensable [36]. In this respect, G2A is different from the other three family members.

Although the ligand–receptor relationship between LPC and G2A is questionable, numerous studies still demonstrate a connection between G2A and LPC and other bioactive lipids [49–59]. G2A is involved in LPC-induced chemotaxis in T lymphocytes (human Jurkat cell line and mouse DO11.10 cell line) and macrophages (mouse peritoneal macrophage and J774A.1 macrophage cell line) [57–59]. Furthermore, for G2A-dependent attraction of phagocytes to apoptotic cells, LPC, but none of the LPC metabolites or other lysoPLs, is essential [55]. G2A is also involved in LPC triggered apoptosis through G_{13} and G_s pathway in HeLa cells (human cervical cancer cell line) [54] and neuritogenesis in rat PC12 cells [52]. In addition to LPC, other lysophospholipids (lyso-PLs) are also showed to stimulate G2A signaling pathways. Lyso-PLs bearing various head groups mobilize neutrophil secretory vesicles in a G2A-dependent manner [50]. Moreover, lysophosphatidylserine (lyso-PS) facilitates the resolution of neutrophilic inflammation [114], and enhances the efferocytosis of dying neutrophils by macrophages via G2A signaling [115]. In addition to lysophospholipids, oxidized free fatty acids are proposed to be another kind of G2A ligands. 9-hydroxyoctadecadienoic acid (HODE) has the highest activity to stimulate calcium responses in G2A-expressing cells. 9-Hydroperoxyoctadecadienoic acid (HPODE) and the arachidonic acid-derived oxidized fatty acids, such as 11-hydroxyeicosatetraenoic acid (HETE), 5-, 8-, 9-, 12-, and 15-HETE also show activity of triggering G2A signaling pathways [116].

Several lines of evidences show that G2A plays important roles in immunity and inflammation. Although G2A is highly expressed in immune cells, G2A-deficient mice do not exhibit any obvious defects in T and B lymphoid development during young adulthood. However, older G2A knockout mice develop a slowly progressive late-onset systemic autoimmune syndrome with enlargement of secondary lymphoid organs and abnormal expansion of lymphocytes [117], suggesting a role of G2A in the regulation of immune system homeostasis. Interestingly, both pro- and antiinflammatory effects of G2A have been reported. For the proinflammatory function, G2A mediates the chemotaxis of lymphocytes and macrophages toward LPC [55, 57–59], a proinflammatory lipid associated with atherosclerosis and other inflammatory conditions [118]. Consistent with these observations, G2A deficiency reduces macrophage accumulation in atherosclerotic lesions and leads to suppression of atherosclerosis in $G2A^{-/-}LDLR^{-/-}$ mice [119]. In addition to lysophospholipids, oxidized free fatty acid, 9-HODE is associated with inflammatory function of G2A. 9-HODE stimulates intracellular calcium mobilization and secretion of proinflammatory cytokines such as IL-6, IL-8, and GM-CSF through G2A receptor in normal human epidermal keratinocytes [120]. These studies suggest a proinflammatory role of G2A. On the other hand, antiinflammatory effects of G2A have also been reported. A study showed that peritoneal macrophages isolated from $G2A^{-/-}ApoE^{-/-}$ mice exhibit increased inflammatory cytokine production and nuclear factor (NF) κB activation compared with that isolated from $G2A^{+/+}ApoE^{-/-}$ mice [121]. Furthermore, G2A is essential for the attraction of macrophages to apoptotic cells such as dying neutrophils to resolve inflammation and prevent autoimmunity [55, 114, 115]. It is also shown that G2A has antiinflammatory function in vascular endothelial cells as the aortas from G2A-deficient mice exhibit higher capacity to adhere to monocytes [122].

Concluding Remarks

Precise regulation of pH homeostasis is crucial for normal physiology. The proton-sensing GPCR family has emerged as important cell-surface receptors for cells to perceive and respond to acidic extracellular pH. Activation of the pH-sensing receptors by acidosis elicits several downstream signaling events such as the G_s/cAMP, G_q/Ca^{2+}/PKC, and G_{13}/Rho pathways and controls many cellular processes. Increasing evidence indicates that these pH-sensing GPCRs play regulatory roles in multiple physiological systems including the cardiovascular, immune, renal, respiratory, skeletal, and nervous systems [20, 21, 29–44].

Acidosis is associated with a variety of diseases such as cancer, inflammation, ischemia, and renal, respiratory, and metabolic diseases [10–16]. The involvement of the pH-sensing GPCRs in the acidosis-related diseases, particularly cancer and inflammation, has been indicated. In cancer, both tumor-promoting and tumor-suppressing effects of the pH-sensing GPCRs have been reported. For example, GPR4 and OGR1 have been shown to inhibit tumor cell migration and metastasis [30, 94, 95], but it has also been shown that GPR4 induces transformation of NIH3T3 immortalized fibroblasts and OGR1 suppresses myeloid cell-mediated antitumor immunity [64, 96]. Similarly, both anti- and protumorigenic activities have been indicated for TDAG8 and G2A [32, 54, 64, 102, 111–113]. These results suggest that the function of the pH-sensing GPCRs in cancer biology is highly dependent on tumor types, cell types, and disease context. In inflammation, the biological roles of the pH-sensing GPCRs are also complex. Upon activation by acidosis in endothelial cells, GPR4 triggers a proinflammatory signal and increases endothelial cell adhesion to leukocytes [20]. On the other hand, acidosis activation of TDAG8 inhibits inflammatory cytokine production in leukocytes [106–108]. Currently, more research is clearly needed to better understand the biology of the pH-sensing GPCRs in physiology and diseases in order to exploit the therapeutic potential of this intriguing receptor family. Furthermore, the differential tissue expression pattern, sequence homology, and divergence of the pH-sensing receptor family members can provide versatile possibilities for modulating the activities of these receptors.

GPCRs are important pharmaceutical targets accounting for 30–50 % of marketed drugs [123−125]. The pH-sensing GPCRs may represent novel targets for the treatment of acidosis-related diseases. Recently, several synthetic small molecule agonists and antagonists have been identified for the pH-sensing GPCRs, including agonists for G2A [126], antagonists for GPR4 [127], an agonist for TDAG8 [108], and an agonist for OGR1 [128]. These small molecules provide useful pharmacological tools to study the function and signaling of the pH-sensing GPCRs. With more understanding of the receptor functions and further development of agonists and antagonists for in vivo applications, the pH-sensing GPCRs may indeed be exploited as useful targets for disease treatment.

Acknowledgments We apologize to the colleagues whose work could not be cited in this manuscript due to space limitation. The research in the authors' laboratory has been supported by grants from the American Heart Association, Brody Brothers Endowment Fund, ECU/Vidant Cancer Research and Education Fund, Golfers against Cancer Foundation, and North Carolina Biotechnology Center (to L.V.Y.). All authors contributed equally to this manuscript. Correspondence should be addressed to L.V.Y. (yangl@ecu.edu).

References

1. Koul PB (2009) Diabetic ketoacidosis: a current appraisal of pathophysiology and management. Clin Pediatr (Phila) 48(2):135–144
2. Kraut JA, Madias NE (2001) Approach to patients with acid-base disorders. Respir Care 46(4):392–403
3. Krieger NS, Frick KK, Bushinsky DA (2004) Mechanism of acid-induced bone resorption. Curr Opin Nephrol Hypertens 13(4):423–436
4. Lemann J Jr, Bushinsky DA, Hamm LL (2003) Bone buffering of acid and base in humans. Am J Physiol Renal Physiol 285(5):F811–F832
5. Aalkjaer C, Peng HL (1997) pH and smooth muscle. Acta Physiol Scand 161(4):557–566
6. De Vito P (2006) The sodium/hydrogen exchanger: a possible mediator of immunity. Cell Immunol 240(2):69–85
7. Fang J, Quinones QJ, Holman TL, Morowitz MJ, Wang Q, Zhao H, Sivo F, Maris JM, Wahl ML (2006) The H^+-linked monocarboxylate transporter (MCT1/SLC16A1): a potential therapeutic target for high-risk neuroblastoma. Mol Pharmacol 70(6):2108–2115
8. Izumi H, Torigoe T, Ishiguchi H, Uramoto H, Yoshida Y, Tanabe M, Ise T, Murakami T, Yoshida T, Nomoto M, Kohno K (2003) Cellular pH regulators: potentially promising molecular targets for cancer chemotherapy. Cancer Treat Rev 29(6):541–549
9. Curley G, Contreras MM, Nichol AD, Higgins BD, Laffey JG (2010) Hypercapnia and acidosis in sepsis: a double-edged sword? Anesthesiology 112(2):462–472
10. Gatenby RA, Gillies RJ (2004) Why do cancers have high aerobic glycolysis? Nat Rev Cancer 4(11):891–899
11. Huang Y, McNamara JO (2004) Ischemic stroke: "acidotoxicity" is a perpetrator. Cell 118(6):665–666
12. Hunt JF, Fang K, Malik R, Snyder A, Malhotra N, Platts-Mills TA, Gaston B (2000) Endogenous airway acidification. Implications for asthma pathophysiology. Am J Respir Crit Care Med 161(3 Pt 1):694–699
13. Kellum JA (2000) Determinants of blood pH in health and disease. Crit Care 4(1):6–14
14. Lardner A (2001) The effects of extracellular pH on immune function. J Leukoc Biol 69(4):522–530

15. Nedergaard M, Kraig RP, Tanabe J, Pulsinelli WA (1991) Dynamics of interstitial and intracellular pH in evolving brain infarct. Am J Physiol 260(3 Pt 2):R581–R588
16. Yang LV, Castellone RD, Dong L (2012) Targeting tumor microenvironments for cancer prevention and therapy. In: Georgakilas AG (ed) Cancer prevention—From mechanisms to translational benefits. InTech, pp 3–40
17. Koppenol WH, Bounds PL, Dang CV (2011) Otto Warburg's contributions to current concepts of cancer metabolism. Nat Rev Cancer 11(5):325–337
18. Helmlinger G, Sckell A, Dellian M, Forbes NS, Jain RK (2002) Acid production in glycolysis-impaired tumors provides new insights into tumor metabolism. Clin Cancer Res 8(4):1284–1291
19. Yamagata M, Hasuda K, Stamato T, Tannock IF (1998) The contribution of lactic acid to acidification of tumours: studies of variant cells lacking lactate dehydrogenase. Br J Cancer 77(11):1726–1731
20. Chen A, Dong L, Leffler NR, Asch AS, Witte ON, Yang LV (2011) Activation of GPR4 by acidosis increases endothelial cell adhesion through the cAMP/Epac pathway. PLoS One 6(11):e27586
21. Huang CW, Tzeng JN, Chen YJ, Tsai WF, Chen CC, Sun WH (2007) Nociceptors of dorsal root ganglion express proton-sensing G-protein-coupled receptors. Mol Cell Neurosci 36(2):195–210
22. Holzer P (2009) Acid-sensitive ion channels and receptors. Handb Exp Pharmacol 194:283–332
23. Smith JB, Dwyer SD, Smith L (1989) Lowering extracellular Ph evokes inositol polyphosphate formation and calcium mobilization. J Biol Chem 264(15):8723–8728
24. Dwyer SD, Zhuang Y, Smith JB (1991) Calcium mobilization by cadmium or decreasing extracellular Na^+ or pH in coronary endothelial cells. Exp Cell Res 192(1):22–31
25. Negulescu PA, Machen TE (1990) Lowering extracellular sodium or pH raises intracellular calcium in gastric cells. J Membr Biol 116(3):239–248
26. Rohra DK, Saito SY, Ohizumi Y (2003) Mechanism of acidic pH-induced contraction in spontaneously hypertensive rat aorta: role of Ca^{2+} release from the sarcoplasmic reticulum. Acta Physiol Scand 179(3):273–280
27. Saadoun S, Lluch M, Rodriguez-Alvarez J, Blanco I, Rodriguez R (1998) Extracellular acidification modifies Ca^{2+} fluxes in rat brain synaptosomes. Biochem Biophys Res Commun 242(1):123–128
28. Trevani AS, Andonegui G, Giordano M, Lopez DH, Gamberale R, Minucci F, Geffner JR (1999) Extracellular acidification induces human neutrophil activation. J Immunol 162(8):4849–4857
29. Ludwig MG, Vanek M, Guerini D, Gasser JA, Jones CE, Junker U, Hofstetter H, Wolf RM, Seuwen K (2003) Proton-sensing G-protein-coupled receptors. Nature 425(6953):93–98
30. Castellone RD, Leffler NR, Dong L, Yang LV (2011) Inhibition of tumor cell migration and metastasis by the proton-sensing GPR4 receptor. Cancer Lett 312:197–208
31. Frick KK, Krieger NS, Nehrke K, Bushinsky DA (2009) Metabolic acidosis increases intracellular calcium in bone cells through activation of the proton receptor OGR1. J Bone Miner Res 24(2):305–313
32. Ihara Y, Kihara Y, Hamano F, Yanagida K, Morishita Y, Kunita A, Yamori T, Fukayama M, Aburatani H, Shimizu T, Ishii S (2010) The G protein-coupled receptor T-cell death-associated gene 8 (TDAG8) facilitates tumor development by serving as an extracellular pH sensor. Proc Natl Acad Sci U S A 107(40):17309–17314
33. Ishii S, Kihara Y, Shimizu T (2005) Identification of T cell death-associated gene 8 (TDAG8) as a novel acid sensing G-protein-coupled receptor. J Biol Chem 280(10):9083–9087
34. Kottyan LC, Collier AR, Cao KH, Niese KA, Hedgebeth M, Radu CG, Witte ON, Khurana Hershey GK, Rothenberg ME, Zimmermann N (2009) Eosinophil viability is increased by acidic pH in a cAMP- and GPR65-dependent manner. Blood 114(13):2774–2782
35. Murakami N, Yokomizo T, Okuno T, Shimizu T (2004) G2A is a proton-sensing G-protein-coupled receptor antagonized by lysophosphatidylcholine. J Biol Chem 279(41):42484–42491

36. Radu CG, Nijagal A, McLaughlin J, Wang L, Witte ON (2005) Differential proton sensitivity of related G protein-coupled receptors T cell death-associated gene 8 and G2A expressed in immune cells. Proc Natl Acad Sci U S A 102(5):1632–1637
37. Seuwen K, Ludwig MG, Wolf RM (2006) Receptors for protons or lipid messengers or both? J Recept Signal Transduct Res 26(5–6):599–610
38. Sun X, Yang LV, Tiegs BC, Arend LJ, McGraw DW, Penn RB, Petrovic S (2010) Deletion of the pH sensor GPR4 decreases renal acid excretion. J Am Soc Nephrol 21(10):1745–1755
39. Wang JQ, Kon J, Mogi C, Tobo M, Damirin A, Sato K, Komachi M, Malchinkhuu E, Murata N, Kimura T, Kuwabara A, Wakamatsu K, Koizumi H, Uede T, Tsujimoto G, Kurose H, Sato T, Harada A, Misawa N, Tomura H, Okajima F (2004) TDAG8 is a proton-sensing and psychosine-sensitive G-protein-coupled receptor. J Biol Chem 279(44):45626–45633
40. Yang LV, Radu CG, Roy M, Lee S, McLaughlin J, Teitell MA, Iruela-Arispe ML, Witte ON (2007) Vascular abnormalities in mice deficient for the G protein-coupled receptor GPR4 that functions as a pH sensor. Mol Cell Biol 27(4):1334–1347
41. Hang LH, Yang JP, Yin W, Wang LN, Guo F, Ji FH, Shao DH, Xu QN, Wang XY, Zuo JL (2012) Activation of spinal TDAG8 and its downstream PKA signaling pathway contribute to bone cancer pain in rats. Eur J Neurosci 36(1):2107–2117
42. Liu JP, Nakakura T, Tomura H, Tobo M, Mogi C, Wang JQ, He XD, Takano M, Damirin A, Komachi M, Sato K, Okajima F (2010) Each one of certain histidine residues in G-protein-coupled receptor GPR4 is critical for extracellular proton-induced stimulation of multiple G-protein-signaling pathways. Pharmacol Res 61(6):499–505
43. Saxena H, Deshpande DA, Tiegs BC, Yan H, Battafarano RJ, Burrows WM, Damera G, Panettieri RA, Dubose TD Jr, An SS, Penn RB (2012) The GPCR OGR1 (GPR68) mediates diverse signalling and contraction of airway smooth muscle in response to small reductions in extracellular pH. Br J Pharmacol 166(3):981–990
44. Tobo M, Tomura H, Mogi C, Wang JQ, Liu JP, Komachi M, Damirin A, Kimura T, Murata N, Kurose H, Sato K, Okajima F (2007) Previously postulated "ligand-independent" signaling of GPR4 is mediated through proton-sensing mechanisms. Cell Signal 19(8):1745–1753
45. Im DS, Heise CE, Nguyen T, O'Dowd BF, Lynch KR (2001) Identification of a molecular target of psychosine and its role in globoid cell formation. J Cell Biol 153(2):429–434
46. Kabarowski JH, Zhu K, Le LQ, Witte ON, Xu Y (2001) Lysophosphatidylcholine as a ligand for the immunoregulatory receptor G2A. Science 293(5530):702–705. (Retraction, 307:206, 2005)
47. Xu Y, Zhu K, Hong G, Wu W, Baudhuin LM, Xiao Y, Damron DS (2000) Sphingosylphosphorylcholine is a ligand for ovarian cancer G-protein-coupled receptor 1. Nat Cell Biol 2(5):261–267. (Retraction, 8:299, 2006)
48. Zhu K, Baudhuin LM, Hong G, Williams FS, Cristina KL, Kabarowski JH, Witte ON, Xu Y (2001) Sphingosylphosphorylcholine and lysophosphatidylcholine are ligands for the G protein-coupled receptor GPR4. J Biol Chem 276(44):41325–41335. (Retraction, 280:43280, 2005)
49. Ding WG, Toyoda F, Ueyama H, Matsuura H (2011) Lysophosphatidylcholine enhances I(Ks) currents in cardiac myocytes through activation of G protein, PKC and Rho signaling pathways. J Mol Cell Cardiol 50(1):58–65
50. Frasch SC, Zemski-Berry K, Murphy RC, Borregaard N, Henson PM, Bratton DL (2007) Lysophospholipids of different classes mobilize neutrophil secretory vesicles and induce redundant signaling through G2A. J Immunol 178(10):6540–6548
51. Hasegawa H, Lei J, Matsumoto T, Onishi S, Suemori K, Yasukawa M (2011) Lysophosphatidylcholine enhances the suppressive function of human naturally occurring regulatory T cells through TGF-beta production. Biochem Biophys Res Commun 415(3):526–531
52. Ikeno Y, Konno N, Cheon SH, Bolchi A, Ottonello S, Kitamoto K, Arioka M (2005) Secretory phospholipases A2 induce neurite outgrowth in PC12 cells through lysophosphatidylcholine generation and activation of G2A receptor. J Biol Chem 280(30):28044–28052
53. Khan SY, McLaughlin NJ, Kelher MR, Eckels P, Gamboni-Robertson F, Banerjee A, Silliman CC (2010) Lysophosphatidylcholines activate G2A inducing $G_{\alpha i-1}$-/$G_{\alpha q/11}$-Ca^{2+} flux, $G_{\beta\gamma}$-Hck activation and clathrin/β-arrestin-1/GRK6 recruitment in PMNs. Biochem J 432(1):35–45

54. Lin P, Ye RD (2003) The lysophospholipid receptor G2A activates a specific combination of G proteins and promotes apoptosis. J Biol Chem 278(16):14379–14386
55. Peter C, Waibel M, Radu CG, Yang LV, Witte ON, Schulze-Osthoff K, Wesselborg S, Lauber K (2008) Migration to apoptotic "find-me" signals is mediated via the phagocyte receptor G2A. J Biol Chem 283(9):5296–5305
56. Qin ZX, Zhu HY, Hu YH (2009) Effects of lysophosphatidylcholine on beta-amyloid-induced neuronal apoptosis. Acta Pharmacol Sin 30(4):388–395
57. Radu CG, Yang LV, Riedinger M, Au M, Witte ON (2004) T cell chemotaxis to lysophosphatidylcholine through the G2A receptor. Proc Natl Acad Sci U S A 101(1):245–250
58. Wang L, Radu CG, Yang LV, Bentolila LA, Riedinger M, Witte ON (2005) Lysophosphatidylcholine-induced surface redistribution regulates signaling of the murine G protein-coupled receptor G2A. Mol Biol Cell 16(5):2234–2247
59. Yang LV, Radu CG, Wang L, Riedinger M, Witte ON (2005) Gi-independent macrophage chemotaxis to lysophosphatidylcholine via the immunoregulatory GPCR G2A. Blood 105(3):1127–1134
60. Mahadevan MS, Baird S, Bailly JE, Shutler GG, Sabourin LA, Tsilfidis C, Neville CE, Narang M, Korneluk RG (1995) Isolation of a novel G protein-coupled receptor (GPR4) localized to chromosome 19q13.3. Genomics 30(1):84–88
61. Lum H, Qiao J, Walter RJ, Huang F, Subbaiah PV, Kim KS, Holian O (2003) Inflammatory stress increases receptor for lysophosphatidylcholine in human microvascular endothelial cells. Am J Physiol Heart Circ Physiol 285(4):H1786–H1789
62. Kim KS, Ren J, Jiang Y, Ebrahem Q, Tipps R, Cristina K, Xiao YJ, Qiao J, Taylor KL, Lum H, Anand-Apte B, Xu Y (2005) GPR4 plays a critical role in endothelial cell function and mediates the effects of sphingosylphosphorylcholine. Faseb J 19(7):819–821
63. Wyder L, Suply T, Ricoux B, Billy E, Schnell C, Baumgarten BU, Maira SM, Koelbing C, Ferretti M, Kinzel B, Muller M, Seuwen K, Ludwig MG (2011) Reduced pathological angiogenesis and tumor growth in mice lacking GPR4, a proton sensing receptor. Angiogenesis 14(4):533–544
64. Sin WC, Zhang Y, Zhong W, Adhikarakunnathu S, Powers S, Hoey T, An S, Yang J (2004) G protein-coupled receptors GPR4 and TDAG8 are oncogenic and overexpressed in human cancers. Oncogene 23(37):6299–6303
65. Bektas M, Barak LS, Jolly PS, Liu H, Lynch KR, Lacana E, Suhr KB, Milstien S, Spiegel S (2003) The G protein-coupled receptor GPR4 suppresses ERK activation in a ligand-independent manner. Biochemistry 42(42):12181–12191
66. Qiao J, Huang F, Naikawadi RP, Kim KS, Said T, Lum H (2006) Lysophosphatidylcholine impairs endothelial barrier function through the G protein-coupled receptor, GPR4. Am J Physiol Lung Cell Mol Physiol 291:L91–L101
67. Huang F, Mehta D, Predescu S, Kim KS, Lum H (2007) A novel lysophospholipid- and pH-sensitive receptor, GPR4, in brain endothelial cells regulates monocyte transmigration. Endothelium 14(1):25–34
68. Zou Y, Kim CH, Chung JH, Kim JY, Chung SW, Kim MK, Im DS, Lee J, Yu BP, Chung HY (2007) Upregulation of endothelial adhesion molecules by lysophosphatidylcholine. Involvement of G protein-coupled receptor GPR4. FEBS J 274(10):2573–2584
69. Codina J, Opyd TS, Powell ZB, Furdui CM, Petrovic S, Penn RB, DuBose TD Jr (2011) pH-dependent regulation of the α-subunit of H^+-K^+-ATPase (HKα2). Am J Physiol Renal Physiol 301(3):F536–F543
70. Zhang Y, Feng Y, Justus CR, Jiang W, Li Z, Lu JQ, Brock RS, McPeek MK, Weidner DA, Yang LV, Hu XH (2012) Comparative study of 3D morphology and functions on genetically engineered mouse melanoma cells. Integr Biol (Camb) 4(11):1428–1436
71. Xu Y, Casey G (1996) Identification of human OGR1, a novel G protein-coupled receptor that maps to chromosome 14. Genomics 35(2):397–402
72. Mogi C, Tomura H, Tobo M, Wang JQ, Damirin A, Kon J, Komachi M, Hashimoto K, Sato K, Okajima F (2005) Sphingosylphosphorylcholine antagonizes proton-sensing ovarian cancer G-protein-coupled receptor 1 (OGR1)-mediated inositol phosphate production and cAMP accumulation. J Pharmacol Sci 99(2):160–167

73. Tomura H, Wang JQ, Komachi M, Damirin A, Mogi C, Tobo M, Kon J, Misawa N, Sato K, Okajima F (2005) Prostaglandin I(2) production and cAMP accumulation in response to acidic extracellular pH through OGR1 in human aortic smooth muscle cells. J Biol Chem 280(41):34458–34464
74. Takayanagi H, Kim S, Koga T, Nishina H, Isshiki M, Yoshida H, Saiura A, Isobe M, Yokochi T, Inoue J, Wagner EF, Mak TW, Kodama T, Taniguchi T (2002) Induction and activation of the transcription factor NFATc1 (NFAT2) integrate RANKL signaling in terminal differentiation of osteoclasts. Dev Cell 3(6):889–901
75. Komarova SV, Pereverzev A, Shum JW, Sims SM, Dixon SJ (2005) Convergent signaling by acidosis and receptor activator of NF-kappaB ligand (RANKL) on the calcium/calcineurin/NFAT pathway in osteoclasts. Proc Natl Acad Sci U S A 102(7):2643–2648
76. Yang M, Mailhot G, Birnbaum MJ, MacKay CA, Mason-Savas A, Odgren PR (2006) Expression of and role for ovarian cancer G-protein-coupled receptor 1 (OGR1) during osteoclastogenesis. J Biol Chem 281(33):23598–23605
77. Pereverzev A, Komarova SV, Korcok J, Armstrong S, Tremblay GB, Dixon SJ, Sims SM (2008) Extracellular acidification enhances osteoclast survival through an NFAT-independent, protein kinase C-dependent pathway. Bone 42(1):150–161
78. Li H, Wang D, Singh LS, Berk M, Tan H, Zhao Z, Steinmetz R, Kirmani K, Wei G, Xu Y (2009) Abnormalities in osteoclastogenesis and decreased tumorigenesis in mice deficient for ovarian cancer G protein-coupled receptor 1. PLoS One 4(5):e5705
79. Tomura H, Wang JQ, Liu JP, Komachi M, Damirin A, Mogi C, Tobo M, Nochi H, Tamoto K, Im DS, Sato K, Okajima F (2008) Cyclooxygenase-2 expression and prostaglandin E2 production in response to acidic pH through OGR1 in a human osteoblastic cell line. J Bone Miner Res 23(7):1129–1139
80. Frick KK, Bushinsky DA (2010) Effect of metabolic and respiratory acidosis on intracellular calcium in osteoblasts. Am J Physiol Renal Physiol 299(2):F418–F425
81. Krieger NS, Bushinsky DA (2011) Pharmacological inhibition of intracellular calcium release blocks acid-induced bone resorption. Am J Physiol Renal Physiol 300(1):F91–F97
82. Brenninkmeijer L, Kuehl C, Geldart AM, Arons E, Christou H (2011) Heme oxygenase-1 does not mediate the effects of extracellular acidosis on vascular smooth muscle cell proliferation, migration, and susceptibility to apoptosis. J Vasc Res 48(4):285–296
83. Ishizaka H, Gudi SR, Frangos JA, Kuo L (1999) Coronary arteriolar dilation to acidosis: role of ATP-sensitive potassium channels and pertussis toxin-sensitive G proteins. Circulation 99(4):558–563
84. Iwasawa K, Nakajima T, Hazama H, Goto A, Shin WS, Toyo-oka T, Omata M (1997) Effects of extracellular pH on receptor-mediated Ca^{2+} influx in A7r5 rat smooth muscle cells: involvement of two different types of channel. J Physiol 503(Pt 2):237–251
85. Leffler CW, Balabanova L, Williams KK (1999) cAMP production by piglet cerebral vascular smooth muscle cells: pH(o), pH(i), and permissive action of PGI(2). Am J Physiol 277(5 Pt 2):H1878–H1883
86. Liu JP, Komachi M, Tomura H, Mogi C, Damirin A, Tobo M, Takano M, Nochi H, Tamoto K, Sato K, Okajima F (2010) Ovarian cancer G protein-coupled receptor 1-dependent and -independent vascular actions to acidic pH in human aortic smooth muscle cells. Am J Physiol Heart Circ Physiol 299(3):H731–H742
87. Kodric M, Shah AN, Fabbri LM, Confalonieri M (2007) An investigation of airway acidification in asthma using induced sputum: a study of feasibility and correlation. Am J Respir Crit Care Med 175(9):905–910
88. Poschet J, Perkett E, Deretic V (2002) Hyperacidification in cystic fibrosis: links with lung disease and new prospects for treatment. Trends Mol Med 8(11):512–519
89. Ricciardolo FL, Gaston B, Hunt J (2004) Acid stress in the pathology of asthma. J Allergy Clin Immunol 113(4):610–619
90. Ichimonji I, Tomura H, Mogi C, Sato K, Aoki H, Hisada T, Dobashi K, Ishizuka T, Mori M, Okajima F (2010) Extracellular acidification stimulates IL-6 production and $Ca(^{2+})$ mobilization through proton-sensing OGR1 receptors in human airway smooth muscle cells. Am J Physiol Lung Cell Mol Physiol 299(4):L567–L577

91. Matsuzaki S, Ishizuka T, Yamada H, Kamide Y, Hisada T, Ichimonji I, Aoki H, Yatomi M, Komachi M, Tsurumaki H, Ono A, Koga Y, Dobashi K, Mogi C, Sato K, Tomura H, Mori M, Okajima F (2011) Extracellular acidification induces connective tissue growth factor production through proton-sensing receptor OGR1 in human airway smooth muscle cells. Biochem Biophys Res Commun 413(4):499–503
92. Liu C, Li Q, Zhou X, Kolosov VP, Perelman JM (2013) Regulator of G-protein signaling 2 inhibits acid-induced mucin5AC hypersecretion in human airway epithelial cells. Respir Physiol Neurobiol 185(2):265–271
93. Huang WC, Swietach P, Vaughan-Jones RD, Ansorge O, Glitsch MD (2008) Extracellular acidification elicits spatially and temporally distinct Ca^{2+} signals. Curr Biol 18(10):781–785
94. Singh LS, Berk M, Oates R, Zhao Z, Tan H, Jiang Y, Zhou A, Kirmani K, Steinmetz R, Lindner D, Xu Y (2007) Ovarian cancer G protein-coupled receptor 1, a new metastasis suppressor gene in prostate cancer. J Natl Cancer Inst 99(17):1313–1327
95. Ren J, Zhang L (2011) Effects of ovarian cancer G protein coupled receptor 1 on the proliferation, migration, and adhesion of human ovarian cancer cells. Chin Med J (Engl) 124(9):1327–1332
96. Yan L, Singh LS, Zhang L, Xu Y (2012) Role of OGR1 in myeloid-derived cells in prostate cancer. Oncogene 33:157–164
97. Mohebbi N, Benabbas C, Vidal S, Daryadel A, Bourgeois S, Velic A, Ludwig MG, Seuwen K, Wagner CA (2012) The proton-activated G protein coupled receptor OGR1 acutely regulates the activity of epithelial proton transport proteins. Cell Physiol Biochem 29(3–4):313–324
98. Nakakura T, Mogi C, Tobo M, Tomura H, Sato K, Kobayashi M, Ohnishi H, Tanaka S, Wayama M, Sugiyama T, Kitamura T, Harada A, Okajima F (2012) Deficiency of proton-sensing ovarian cancer G protein-coupled receptor 1 attenuates glucose-stimulated insulin secretion. Endocrinology 153(9):4171–4180
99. Choi JW, Lee SY, Choi Y (1996) Identification of a putative G protein-coupled receptor induced during activation-induced apoptosis of T cells. Cell Immunol 168(1):78–84
100. Kyaw H, Zeng Z, Su K, Fan P, Shell BK, Carter KC, Li Y (1998) Cloning, characterization, and mapping of human homolog of mouse T-cell death-associated gene. DNA Cell Biol 17(6):493–500
101. He XD, Tobo M, Mogi C, Nakakura T, Komachi M, Murata N, Takano M, Tomura H, Sato K, Okajima F (2011) Involvement of proton-sensing receptor TDAG8 in the anti-inflammatory actions of dexamethasone in peritoneal macrophages. Biochem Biophys Res Commun 415(4):627–631
102. Malone MH, Wang Z, Distelhorst CW (2004) The glucocorticoid-induced gene tdag8 encodes a pro-apoptotic G protein-coupled receptor whose activation promotes glucocorticoid-induced apoptosis. J Biol Chem 279(51):52850–52859
103. Radu CG, Cheng D, Nijagal A, Riedinger M, McLaughlin J, Yang LV, Johnson J, Witte ON (2006) Normal immune development and glucocorticoid-induced thymocyte apoptosis in mice deficient for the T-cell death-associated gene 8 receptor. Mol Cell Biol 26(2):668–677
104. Tosa N, Murakami M, Jia WY, Yokoyama M, Masunaga T, Iwabuchi C, Inobe M, Iwabuchi K, Miyazaki T, Onoe K, Iwata M, Uede T (2003) Critical function of T cell death-associated gene 8 in glucocorticoid-induced thymocyte apoptosis. Int Immunol 15(6):741–749
105. McGuire J, Herman JP, Ghosal S, Eaton K, Sallee FR, Sah R (2009) Acid-sensing by the T cell death-associated gene 8 (TDAG8) receptor cloned from rat brain. Biochem Biophys Res Commun 386(3):420–425
106. Mogi C, Tobo M, Tomura H, Murata N, He XD, Sato K, Kimura T, Ishizuka T, Sasaki T, Sato T, Kihara Y, Ishii S, Harada A, Okajima F (2009) Involvement of proton-sensing TDAG8 in extracellular acidification-induced inhibition of proinflammatory cytokine production in peritoneal macrophages. J Immunol 182(5):3243–3251
107. Onozawa Y, Komai T, Oda T (2011) Activation of T cell death-associated gene 8 attenuates inflammation by negatively regulating the function of inflammatory cells. Eur J Pharmacol 654(3):315–319

108. Onozawa Y, Fujita Y, Kuwabara H, Nagasaki M, Komai T, Oda T (2012) Activation of T cell death-associated gene 8 regulates the cytokine production of T cells and macrophages in vitro. Eur J Pharmacol 683(1–3):325–331
109. Chen YJ, Huang CW, Lin CS, Chang WH, Sun WH (2009) Expression and function of proton-sensing G-protein-coupled receptors in inflammatory pain. Mol Pain 5:39
110. Ryder C, McColl K, Zhong F, Distelhorst CW (2012) Acidosis promotes Bcl-2 family-mediated evasion of apoptosis: involvement of acid-sensing G protein-coupled receptor Gpr65 signaling to Mek/Erk. J Biol Chem 287(33):27863–27875
111. Weng Z, Fluckiger AC, Nisitani S, Wahl MI, Le LQ, Hunter CA, Fernal AA, Le Beau MM, Witte ON (1998) A DNA damage and stress inducible G protein-coupled receptor blocks cells in G2/M. Proc Natl Acad Sci U S A 95(21):12334–12339
112. Le LQ, Kabarowski JH, Wong S, Nguyen K, Gambhir SS, Witte ON (2002) Positron emission tomography imaging analysis of G2A as a negative modifier of lymphoid leukemogenesis initiated by the BCR-ABL oncogene. Cancer Cell 1(4):381–391
113. Zohn IE, Klinger M, Karp X, Kirk H, Symons M, Chrzanowska-Wodnicka M, Der CJ, Kay RJ (2000) G2A is an oncogenic G protein-coupled receptor. Oncogene 19(34):3866–3877
114. Frasch SC, Berry KZ, Fernandez-Boyanapalli R, Jin HS, Leslie C, Henson PM, Murphy RC, Bratton DL (2008) NADPH oxidase-dependent generation of lysophosphatidylserine enhances clearance of activated and dying neutrophils via G2A. J Biol Chem 283(48):33736–33749
115. Frasch SC, Fernandez-Boyanapalli RF, Berry KZ, Leslie CC, Bonventre JV, Murphy RC, Henson PM, Bratton DL (2011) Signaling via macrophage G2A enhances efferocytosis of dying neutrophils by augmentation of Rac activity. J Biol Chem 286(14):12108–12122
116. Obinata H, Hattori T, Nakane S, Tatei K, Izumi T (2005) Identification of 9-hydroxyoctadecadienoic acid and other oxidized free fatty acids as ligands of the G protein-coupled receptor G2A. J Biol Chem 280(49):40676–40683
117. Le LQ, Kabarowski JH, Weng Z, Satterthwaite AB, Harvill ET, Jensen ER, Miller JF, Witte ON (2001) Mice lacking the orphan G protein-coupled receptor G2A develop a late-onset autoimmune syndrome. Immunity 14(5):561–571
118. Matsumoto T, Kobayashi T, Kamata K (2007) Role of lysophosphatidylcholine (LPC) in atherosclerosis. Curr Med Chem 14(30):3209–3220
119. Parks BW, Lusis AJ, Kabarowski JH (2006) Loss of the lysophosphatidylcholine effector, G2A, ameliorates aortic atherosclerosis in low-density lipoprotein receptor knockout mice. Arterioscler Thromb Vasc Biol 26(12):2703–2709
120. Hattori T, Obinata H, Ogawa A, Kishi M, Tatei K, Ishikawa O, Izumi T (2008) G2A plays proinflammatory roles in human keratinocytes under oxidative stress as a receptor for 9-hydroxyoctadecadienoic acid. J Invest Dermatol 128(5):1123–1133
121. Bolick DT, Skaflen MD, Johnson LE, Kwon SC, Howatt D, Daugherty A, Ravichandran KS, Hedrick CC (2009) G2A deficiency in mice promotes macrophage activation and atherosclerosis. Circ Res 104(3):318–327
122. Bolick DT, Whetzel AM, Skaflen M, Deem TL, Lee J, Hedrick CC (2007) Absence of the G protein-coupled receptor G2A in mice promotes monocyte/endothelial interactions in aorta. Circ Res 100(4):572–580
123. Jacoby E, Bouhelal R, Gerspacher M, Seuwen K (2006) The 7 TM G-protein-coupled receptor target family. ChemMedChem 1(8):761–782
124. Lappano R, Maggiolini M (2011) G protein-coupled receptors: novel targets for drug discovery in cancer. Nat Rev Drug Discov 10(1):47–60
125. Overington JP, Al-Lazikani B, Hopkins AL (2006) How many drug targets are there? Nat Rev Drug Discov 5(12):993–996
126. Bercher M, Hanson B, van Staden C, Wu K, Ng GY, Lee PH (2009) Agonists of the orphan human G2A receptor identified from inducible G2A expression and beta-lactamase reporter screen. Assay Drug Dev Technol 7(2):133–142

127. Taracido IC, Harrington EM, Hersperger R, Lattmann R, Miltz W, Weigand K (2009) Imidazo pyridine derivatives. United States patent application No. 12/468,706, pp 1–50
128. Russell JL, Goetsch SC, Aguilar HR, Coe H, Luo X, Liu N, van Rooij E, Frantz DE, Schneider JW (2012) Regulated expression of pH sensing G Protein-coupled receptor-68 identified through chemical biology defines a new drug target for ischemic heart disease. ACS Chem Biol 7(6):1077–1083

Part II
“Response to Acidity”

Chapter 5
Response to Acidity: The MondoA–TXNIP Checkpoint Couples the Acidic Tumor Microenvironment to Cell Metabolism

Zhizhou Ye and Donald E. Ayer

List of Abbreviations

α-KG	α-ketoglutarate
AMPK	AMP-activated protein kinase
AOA	aminooxyacetate
ARRDC4	arrestin domain containing 4
B-ALL	B cell acute lymphoblastic leukemia
bHLHZip	basic helix-loop-helix leucine zipper
ChREBP	carbohydrate-response element-binding protein
DCD	dimerization and cytoplasmic localization domain
ECM	extracellular matrix
G6P	glucose 6-phosphate
GPCR	G protein-coupled receptor
GRE	glucocorticoid response element
HDAC	histone deacetylase
HIF-1	hypoxia-inducible factor 1
HMECs	human mammary epithelial cells
LA	lactic acidosis
LDHA	lactate dehydrogenase A
MCR	mondo conserved region
MCT	monocarboxylate/H^+ cotransporter
MEFs	murine embryonic fibroblasts
Mlx	max-like-protein X
mTORC1	mammalian target of rapamycin complex 1
NES	nuclear export signal
NHE	Na^+/H^+ exchanger
OMM	outer membrane of the mitochondria
OXPHOS	oxidative phosphorylation

D. E. Ayer (✉) · Z. Ye
Huntsman Cancer Institute, Department of Oncological Sciences,
University of Utah, 2000 Circle of Hope, Salt Lake City, UT 84112-5550, USA
e-mail: don.ayer@hci.utah.edu

J-T. A. Chi (ed.), *Molecular Genetics of Dysregulated pH Homeostasis,*
DOI 10.1007/978-1-4939-1683-2_5

PFK1	prosphofructokinase 1
PHD2	prolyl hydroxylase domain 2
pHe	extracellular pH
pHi	intracellular pH
ROS	reactive oxygen species
TAD	transcriptional activation domain
TCA	tricarboxylic acid
TME	Tumor Microenvironment
TXNIP	thioredoxin-interacting protein
v-ATPase	vacuolar-type H^+-ATPase

Introduction

Non-transformed cells control their growth and division tightly with checkpoints in place that sense and respond to a variety of stresses. When triggered, these checkpoints halt or slow growth until homeostasis can be restored, or trigger apoptosis when the stress cannot be alleviated. p53 is a quintessential checkpoint protein that allows cells to respond to many different stresses, including DNA damage and defects in ribosomal biogenesis. p53 can also be activated by reduced intracellular bioenergetic charge following activation by adenosine monophosphate (AMP)-activated protein kinase (AMPK) [62]. Underscoring the essential nature of p53 as a checkpoint, its gene TP53 is mutated in a broad spectrum of cancers and cells with mutant p53 push through checkpoints, avoiding cell cycle arrest, senescence, and apoptosis [100].

Like mutation of TP53, another common feature of cancer cells is a reprogramming of their metabolism to one that favors high aerobic glycolysis, yet compared to our understanding of other checkpoint pathways, the mechanisms by which cells sense and respond to metabolic reprogramming are only just now coming into focus [154, 159]. A key player in this regard is the MondoA transcription factor that senses high glycolytic flux and restricts glucose uptake and aerobic glycolysis via its upregulation of the thioredoxin-interacting protein (TXNIP) and its paralog, arrestin domain containing 4 (ARRDC4) [63, 115, 144, 145];. Lactate is produced in the final step of the glycolytic pathway and is exported from the cell along with a proton. Thus, cancer cells with elevated aerobic glycolysis acidify their growth medium, and cancers typically drive lactic acidosis (LA) in their microenvironment. Recent experiments from the Chi lab show that LA drives the nuclear activity of MondoA at the TXNIP and ARRDC4 promoters [17]. In this review, we present evidence and the concept that the regulation of TXNIP and ARRDC4 by MondoA in response to elevated aerobic glycolysis and LA represents a bioenergetic checkpoint and microenvironmental checkpoint. Further, we suggest that common oncogenic lesions inactivate the MondoA–TXNIP/ARRDC4 checkpoint, perhaps contributing to or being permissive for tumor progression.

Cancer Metabolism: A 10,000-Foot View

Glucose and glutamine are essential nutrients required to support the biosynthesis of macromolecules and cofactors, e.g., nucleotides, lipids, amino acids, and nicotinamide adenine dinucleotide phosphate (NADPH), required for cell growth and division. In normal cells, macromolecular biosynthesis is under tight regulatory control, yet anabolic pathways are often dysregulated in cancer cells. The resulting metabolic reprogramming in cancer cells, simplistically high rates of glycolysis and glutaminolysis, supports their high growth and division rates. The mechanisms that underlie the metabolic changes in cancer are at least partially understood, providing a rationale for their targeting with known or novel therapeutics. A number of excellent recent review articles highlight dysregulated metabolic pathways in cancer [23, 24, 150]. Our group is focused on how cells sense and respond transcriptionally to glucose and glutamine and how an extended family of Myc-related transcription factors contributes to metabolic homeostasis in normal and neoplastic cells [64, 114, 138].

Elevated rates of glycolysis and glutaminolysis provide cancer cells, both cell autonomous and nonautonomous advantages. From a cell autonomous perspective, a high rate of aerobic glycolysis provides adequate adenosine triphosphate (ATP) to meet the cell's bioenergetic needs, provides carbon backbones for biosynthetic reactions, and can activate pro-survival pathways [138]. Similarly, glutamine and high glutaminolysis have pleiotropic functions within cancer cells. For example, high glutaminolysis can fill the tricarboxylic acid (TCA) cycle, generate NADPH to support biosynthetic reactions, increase glutathione levels to buffer intracellular redox status, provide nitrogen for base synthesis, and support the synthesis of multiple amino acids [22, 132]. One dominating cell-nonautonomous feature driven by the high rates of glycolysis and glutaminolysis is an acidic tumor microenvironment. Lactate is the terminal product of glycolysis, and a significant portion of glutamine can also be converted to lactate following several steps. Lactate is cotransported along with a proton across the plasma membrane into the extracellular milieu via a variety of acid transporters, generating LA outside of the cell. The glutamine to glutamate conversion also releases an ammonium ion, which can induce autophagy intracellularly and potentially in the neighboring stromal cells in the context of the tumor microenvironment [32]. As discussed in the following sections, the LA microenvironment provides the tumor cell with a number of cell-nonautonomous advantages, and emerging evidence suggests that it may also allow for metabolic coupling between tumor and stromal compartments.

Overview of the Acidic Microenvironment

The tumor microenvironment refers to the host elements surrounding the solid tumor. Anatomically, it consists of the extracellular matrix (ECM) and a number of different resident cell types, including fibroblasts, endothelial cells, adipocytes, and immune cells [66, 69, 129, 146]. The tumor microenvironment is characterized by hypoxia, acidosis, accumulation of lactic acid, nutrient depletion (e.g., glucose, glutamine, and growth factors), and other biochemical and metabolic alterations. These stresses are caused by a combination of poor tissue perfusion, abnormal tumor vasculature, uncontrolled proliferation, and dysregulated metabolism of cancer cells during tumor development and progression [44, 45, 151].

Acidosis and lactosis (collectively called LA) is a characteristic feature of the tumor microenvironment. The extracellular pH (pHe) of human and animal tumors is consistently acidic and can reach pH values approaching 6.0 [46, 124]. Tumor lactate levels vary from 7.3 to 25.9 μmol/g [121]. LA stems from the enhanced production of protons and lactate by the tumor cells as a consequence of the adapted anaerobic metabolism to the hypoxic microenvironment or an elevated level of aerobic glycolysis [35, 44]. LA, through both cell-autonomous and nonautonomous mechanisms, enables cancer progression by promoting cell proliferation and apoptosis evasion. Further, LA creates a microenvironment that is optimal for tumor invasion and metastasis (as discussed below).

Initially, decreased pHe of the microenvironment results in a decrease in intracellular pH (pHi), providing a cellular cytotoxic stimulus. As an adaptive and compensatory mechanism, cancer cells restore pH homeostasis by upregulating the expression and/or the activity of multiple pH-regulating transporters. These pH regulators include the Na^+/H^+ exchangers (NHEs), vacuolar-type H^+-ATPase (v-ATPase), monocarboxylate/H^+ cotransporters (MCTs), and $Na^+/HCO3^-$ cotransporters. These transporters serve to extrude the intracellular proton into the extracellular space [13,18, 34]. Consequently, cancer cells have a reversed pH gradient (pHi ~7.2–7.7 and pHe ~6.2–6.8) compared with normal cells (pHi ~6.9–7.1 and pHe ~7.2–7.4) [13]. Below we list several mechanisms by which altered pH homeostasis confers survival and growth advantages to cancer cells. A number of recent reviews cover these mechanisms in more detail, and we only summarize here [10, 13, 142, 158].

Apoptosis

Altered pH homeostasis can affect apoptosis in at least two ways. First, exposure of normal cells to an acidic environment or reduction of pHi via targeted inhibition of NHE drives apoptosis through p53- and caspase-3-dependent mechanisms [107, 126, 161]. In vitro, low pH promotes pore formation by the proapoptotic B cell lymphoma 2 (Bcl-2) family proteins, the cytochrome c-induced assembly of the apoptotic protease activating factor 1 (Apaf-1) apoptosome, and caspase activation (reviewed in [90]). The strong apoptotic drive provided by low pHe allows for the selection and clonal evolution of cells with mutations that block cell death, e.g., inactivation of p53. Second, low pHi is also required for apoptosis driven by other signals in the microenvironment (e.g., nutrient depletion and oxidative stress). Thus, the high pHi in tumor cells renders them resistant to multiple microenvironmental stresses [41, 49, 80, 113].

Growth and Division

The increased pHi in cancer cells stimulates glycolysis by increasing the enzymatic activity of several key glycolytic enzymes, such as phosphofructokinase 1 (PFK1) and lactate dehydrogenase A (LDHA), whose activities are considerably higher at pH 7.5 than at lower pH [1, 40, 73]. Thus, high pHi in cancer cells is a component of a feed-forward mechanism that drives higher levels of aerobic glycolysis, helping them satisfy their high demand for energy and biomass synthesis. In addition, alkalinization of the cytosol is required for the G1 transition and allows cells to evade a number of cell cycle checkpoints [80, 108, 118, 119].

Migration and Invasion

In order to metastasize, tumor cells must migrate and invade the basement membrane. The migratory activity of tumor cells is positively regulated by increased pHi and decreased pHe (reviewed in [158]). Increased pHi at the migrating front facilitates the directed cell migration via several mechanisms, including increasing cell division control protein 42 (CDC42)-mediated cell polarity, promoting the de novo assembly of actin filaments that drive membrane protrusion in migrating cells, and increasing focal adhesion turnover rate [38, 39, 98, 141]. Decreased pHe promotes cell migration, by regulating the dynamics of integrin–ECM attachments [143]. A decreased pHe also enhances cell invasion by activating the cathepsin- and matrix metalloproteinase-mediated degradation of the ECM [28, 81, 96]. Supporting this acid-mediated invasion hypothesis, the Gillies' group recently showed that the regions of highest tumor invasion corresponded to areas of lowest pHe [33].

Therapeutic Implications

In addition to affecting the tumor and stromal cell populations, tumor acidosis may affect the efficacy of many cancer therapeutics. For example, many commonly used cancer drugs (e.g., doxorubicin pKa ~ 8.3, mitoxantrone pKa ~ 8.1) are weak bases, which are protonated and positively charged in the acidic extracellular space and thus have very low rates of passive diffusion across the plasma membrane, which reduces accumulation in the target transformed cell. Conversely, once the weak base drug enters the cancer cell, it becomes deprotonated and uncharged in the alkalinized cytosol, allowing diffusion across the plasma membrane out of the cell. Thus, the high pHi of cancer cell reduces the effective intracellular concentration of weak base chemotherapeutics. By contrast, in non-transformed cells, the situation is reversed, where weak base drugs are preferentially retained inside the cell. This potentially results in toxic nonspecific effects in non-transformed cells [123].

Metabolic Symbiosis Driven by lactic acidosis (LA)

Besides elevated H^+, increased lactate in the tumor microenvironment is another consequence of aerobic glycolysis. Compared to its role in the genesis of the acidic microenvironment, other biological contributions of lactate, which is generally considered the "waste" from glycolysis, are not well studied. However, lactate accumulation in human tumors is associated with metastasis, tumor recurrence, and poor survival [11, 121, 152, 153]. Further, lactate can fuel oxidative metabolism in oxygenated cancer cells close to the blood supply. Specifically, hypoxic cancer cells distal to the blood vessel rely on glycolysis for energy production. Lactate released from glycolytic cancer cells diffuses down its concentration gradient and is actively taken up and utilized by oxygenated cancer cells to produce energy through oxidative phosphorylation (OXPHOS). Importantly, blockade of the lactate importer monocarboxylate transporter 1 (MCT1), which is expressed preferentially on the oxygenated cancer cells, causes tumor growth retardation [139]. Thus, metabolic coupling may be advantageous for the overall fitness of the cancer cell population. Collectively, these data suggest that oxygenated cancer cells use lactate preferentially, reserving glucose for the hypoxic cancer cells, which are incapable of aerobic respiration. This type of metabolic coupling is not restricted to cancer cells, as it is also observed in normal physiological contexts. For example, glycolytic astrocytes can provide lactate for oxidative neurons [54].

More recently, exogenous lactate was shown to induce the stabilization of hypoxia-inducible factor-1α (HIF-1α) protein in oxidative cancer cells under normoxic conditions [21]. This HIF-1α stabilization requires lactate uptake into the oxidative cancer cells, and is mediated by lactate-derived pyruvate, which acts as a competitive inhibitor of α-ketoglutarate (KG)-dependent dioxygenase prolyl hydroxylase domain 2 (PHD2). The authors further confirmed in vivo that lactate triggers tumor growth and angiogenesis, which is abrogated upon MCT1 inhibition. In a similar study, lactate induces HIF-1α stabilization in nonmalignant endothelial cells and leads to HIF-1-dependent angiogenesis [140]. These findings point to a feed-forward mechanism of lactate in facilitating the glycolytic switch of cancer cells in adaptation to hypoxia.

Another way that microenvironmental lactate benefits the cancer cells involves the interplay between the cancer cells and the non-transformed stromal cells, a phenomenon termed the "reverse Warburg effect" [95, 111]. In this model, the cancer cells induce oxidative stress in surrounding stromal fibroblasts, which leads to the autophagic destruction of their mitochondria [86, 87, 88]. Further, the ammonium ion generated from deamination of glutamine during glutaminolysis in cancer cells provides an additional autophagy signal to stromal cells [89, 112]. As a consequence, the stromal cells have a reduced number of functional mitochondria and must undergo a switch to glycolysis to meet their bioenergetic needs. The neo-glycolytic phenotype of stromal cells generates energy-rich metabolites such as lactate, glutamine, and ketones, which in turn feed the adjacent cancer cells [71, 86, 112]. The ability of cancer cells to modulate the metabolism of neighboring stromal cells helps support their growth in the stringent tumor microenvironment. As with metabolic coupling, lactate is an important mediator of this parasitic strategy used by cancer cells.

These two models, metabolic coupling and the reverse Warburg effect, have been established primarily using cell lines and tissue culture models. How broadly applicable these models are to human cancers remains to be elucidated. Nonetheless, these models provide a useful framework for the metabolic interactions between hypoxic and oxygenated tumor cells and between tumor and stromal cells.

MondoA: The Basics

Over the past decade, several of the mechanisms that lead to an elevated rate of aerobic glycolysis in cancer cells have been uncovered. These include activation of oncogenic signaling pathways and the inhibition of tumor suppressor pathways. In many cases, the effects of these changes on elevating aerobic glycolysis are direct, leading to the hypothesis that metabolic changes are a primary, rather than secondary, effect of cellular transformation. Two transcriptional regulators play key roles in driving aerobic glycolysis: the Myc family (c-, N-, and L-) and the HIF family [47, 48]. Myc and HIF-1α drive aerobic glycolysis by activating the expression of most, if not all, of the genes encoding glycolytic enzymes and upregulating expression of glucose transporters. HIF-1α also restricts flux of pyruvate into the mitochondria by upregulating pyruvate dehydrogenase kinase [70]. Myc proteins can also fuel cellular biosynthesis by driving the upregulation of glutamine transporters and glutaminase, i.e., glutaminolysis [43, 120, 162, 165]. Collectively, we know a great deal about cells upregulate aerobic glycolysis, yet our knowledge of how cells sense flux through different metabolic pathways is less developed. Our work suggests an important role for the MondoA transcription factor in monitoring rates of glycolysis and glutaminolysis and coordinating the use of glucose and glutamine.

Like Myc, MondoA is a member of the basic helix-loop-helix leucine zipper (bHLHZip) family of transcription factors [64, 114, 138]. MondoA interacts with another bHLHZip protein called Mlx (Max-like protein X), and MondoA:Mlx heterodimers, following regulated nuclear entry, are capable of binding promoters of target genes and activating their expression. At 919 amino acids, MondoA is among the largest bHLHZip proteins. Mlx, as the name implies, is more Max-like in size, and a 244 amino acid open reading frame encodes the β-isoform. MondoA and Mlx interact through their HLHZip domains. Compared to Max, which forms partnerships with all members of the Myc family, the Mxd family and the Mnt, Mlx has restricted protein partnerships (Fig. 5.1a). Mlx can interact with, MondoA, carbohydrate-response element-binding protein (ChREBP), Mxd1, and Mxd4; ChREBP is a MondoA paralog [6, 7, 94]. The shared binding of Max and Mlx for Mxd1 and Mxd4 provides a physical connection and the opportunity for functional cross talk between Max-centered and Mlx-centered transcriptional networks. This hypothesis has not yet been thoroughly explored.

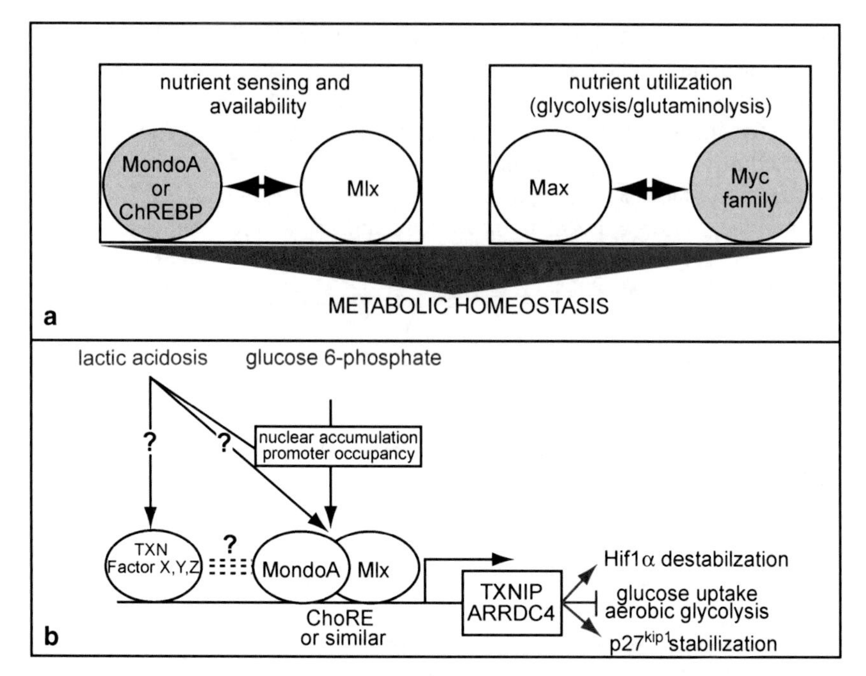

Fig. 5.1 The Mlx network. **a** Current data suggest the existence of Max- and Mlx-centered transcriptional networks that collaborate to control metabolic homeostasis. The Max-centered network controls primarily nutrient utilization by activating or repressing genes involved in glucose and glutamine metabolism. One function of the Mlx-centered network is to control glucose availability. The Max- and Mlx-centered networks may be linked through shared binding of Max and Mlx for Mxd1 and Mxd4 (not shown). **b** MondoA:Mlx complexes activate expression of TXNIP and ARRDC4 via a mechanism that requires G6P for nuclear accumulation, promoter binding, and transcriptional activation. Promoter binding is enhanced by LA by, as indicated by the question marks, unknown mechanisms

In addition to its bHLHZip domain, MondoA has a number of other conserved functional domains. The dimerization and cytoplasmic localization domain (DCD) follows the leucine zipper of both MondoA and Mlx, extending their dimerization interface [30]. The DCD functions as a cytoplasmic anchor, using an uncharacterized mechanism to restrict the nuclear accumulation of monomeric MondoA or Mlx. The cytoplasmic anchoring activity of the DCD is abrogated when MondoA and Mlx interact, creating a dimer that is permissive for nuclear entry and accumulation. MondoA has a potent transcriptional activation domain (TAD) localized in the middle third of its open reading frame, but like Max, Mlx appears to lack a TAD [6–8]. The amino-terminus of the Mondo family has five regions of conservation, which we called the Mondo conserved regions (MCRs) [6, 30]. The MCRs are highly conserved across metazoans and are easily identifiable in lower eukaryotes, suggesting an important ancestral function. Nuclear accumulation, promoter binding, and transcriptional activity of MondoA:Mlx complexes are controlled by glucose 6-phosphate (G6P) [115]. These activities all map to the MCRs.

We believe that G6P interacts directly with the MCRs to control the nuclear activity of MondoA:Mlx complexes by an allosteric mechanism. Supporting this, a recent phylogenetic study of MondoA's amino-terminus proposed the existence MCR6 that can be modeled as a glucose-binding domain [93].

MondoA and ChREBP are paralogs sharing extensive homology in their MCR (52 % identical), bHLHZip (61 % identical), and DCD (70 % identical) domains [92]. Like MondoA, ChREBP also functions as a glucose sensor and interacts with Mlx [55]. Here, we limit our discussion to the functions of MondoA's MCRs, although the analogous domains in ChREBP seem to function similarly. Initially, MondoA and ChREBP were found to be most highly expressed in skeletal muscle and liver, respectively [8, 149]. Given the roles of these tissues in glucose homeostasis, their expression pattern makes sense from a physiological perspective. However, both MondoA and ChREBP are broadly expressed (www.biogps.org) and likely have uncharacterized functions outside of controlling glucose homeostasis.

Unlike other members of the Myc, Max and Mxd network, MondoA:Mlx complexes are not constitutively nuclear [8]. Rather they localize to the outer membrane of the mitochondria (OMM) and shuttle between the OMM and the nucleus [128]. Thus, subcellular localization, rather than expression level, regulates the nuclear and transcriptional activity of MondoA:Mlx complexes. Given the predominance of mitochondria in bioenergetics, we proposed that MondoA:Mlx complexes sense intracellular energy charge at the OMM and communicate that information to the nucleus to trigger an adaptive transcriptional response [128]. We subsequently showed MondoA:Mlx complexes sense glucose and other hexoses and are the predominant regulators of glucose-induced transcription in a breast epithelial cell line [144] and in murine embryonic fibroblasts (MEFs; our unpublished data). ChREBP, which is also held latently in the cytoplasm but not at the OMM [117], plays a similarly predominant role in controlling the glucose-induced transcription in rat hepatocytes [85].

How glucose controls the nuclear accumulation and transcriptional activity of MondoA:Mlx and ChREBP:Mlx complexes is quite complicated, and no unifying model has yet emerged. Early models of ChREBP regulation implicated a number of glucose-dependent phosphorylation and dephosphorylation events in controlling its nuclear activity [149]. The phospho-acceptor sites identified in ChREBP are not conserved in MondoA, suggesting that the regulation of MondoA:Mlx complexes may be simpler than or unrelated to that of ChREBP. Several lines of evidence support the model that MondoA:Mlx complexes sense G6P and more recently fructose 2,6-bisphosphate [115, 116, 144]. Interestingly, the nature of the signaling sugar-derived metabolite may dictate target gene selection. More recent papers have also proposed a more direct role for G6P in regulating ChREBP [25, 78, 79]. Determining which metabolites regulate MondoA:Mlx complexes directly will likely require deciphering the mechanistic details in vitro.

Regardless of the exact mechanism(s) by which glucose regulates MondoA:Mlx nuclear accumulation, its target is almost certainly the MCRs. For example, deletion of the MCRs renders MondoA:Mlx complexes constitutively nuclear, transcriptionally active and independent of glucose [144]. Point mutations in highly conserved residues in any individual MCR fail to phenocopy the deletion of all five MCRs [115], suggesting that the MCRs function as a structural unit. The function of the individual MCRs remains murky. MCRII is a CRM1-dependent nuclear export signal (NES) and MCRIII binds 14-3-3 isoforms [30]. Both MCRII and MCRIII participate in regulating the subcellular localization of MondoA:Mlx complexes [115]. Mutation of MCRII results in the constitutive nuclear localization of MondoA, yet these mutants fail to bind target gene promoters and activate gene expression in the absence of glucose [115]. Thus, nuclear accumulation of MondoA:Mlx complexes is necessary but not sufficient to activate gene expression. Whether MCRI, MCRIV, or MCRV forms additional protein partnerships is not known; however, it is possible that they participate in intramolecular interactions that stabilize structure of the entire MCR domain.

The MondoA–TXNIP Checkpoint

TXNIP is the best-characterized glucose-induced and direct transcriptional target of MondoA:Mlx complexes [144]. As its name indicates, TXNIP was identified as a binding partner and negative regulator of thioredoxin. TXNIP has pleiotropic function; most prominent among these is regulating reactive oxygen species (ROS) by blocking thioredoxin function and restricting glucose uptake and aerobic glycolysis. Recently, TXNIP was shown to be a functional component of the NLRP inflammasome required for the cleavage and production of IL1-α [131]. TXNIP may be a tumor suppressor in a number of cancers, reflecting its proapoptotic activity, its stabilization of the cell cycle inhibitor p27^{Kip1} [60], its destabilization of HIF-1α [137], or its blockade of aerobic glycolysis [109, 110]. In no case has the essential effector function of TXNIP been characterized in detail. TXNIP also controls peripheral glucose uptake into skeletal muscle [106], hepatic glucose production [19], and glucose toxicity of pancreatic β-cells [15] and, by doing so, contributes to organismal glucose homeostasis. Several mouse models of TXNIP loss support this contention. A complete accounting of TXNIP's functions is not possible here, although several recent reviews are available [109, 166, 167].

Because TXNIP can block glucose uptake and aerobic glycolysis, its glucose-dependent induction by MondoA:Mlx complexes drives a negative feedback loop that restricts glucose uptake when glycolytic flux is high (Fig. 5.1b). The TXNIP paralog, ARRDC4, but not other members of this class of arrestin proteins, also restricts glucose uptake [110]. ARRDC4 expression is also induced by MondoA in a glucose-dependent manner [115], suggesting that TXNIP and ARRDC4 function in concert or redundantly downstream of MondoA:Mlx complexes to restrict glucose availability. How TXNIP and ARRDC4 restrict glucose uptake is not fully resolved; however, a recent paper shows that TXNIP can target glucose transporter 1 (GLUT1) for degradation [163].

Current data support a physiological role for the MondoA:Mlx–TXNIP negative feedback loop in blocking aerobic glycolysis, reversing metabolic reprogramming associated with the transformed state. For example, reduction of MondoA or TXNIP levels by knockdown or genetic deletion triggers elevated glucose uptake and aerobic glycolysis [63, 115, 144]. Conversely, expression of a dominant active allele of MondoA potently blocks glucose uptake. This reduction of glucose uptake is partially dependent on TXNIP, so other MondoA targets must also be required [144]. ARRDC4 is one obvious candidate. Finally, MondoA and TXNIP are required to restrict glucose uptake following exposure to different hexose sugars [145], i.e., the hexose curb. The MondoA–TXNIP regulatory circuit is extant in many cell types; however, MondoA appears to have growth promoting and pro-glycolytic functions in B cell acute lymphocytic leukemia (see below).

Glutamine-Dependent Repression of TXNIP

Like glucose, glutamine is also typically required for cell growth. Also like glucose, glutamine can flux into a number of metabolic or biosynthetic pathways [22, 132]. Surprisingly, glucose-stimulated TXNIP induction is repressed by glutamine in some cell types [63]. Glutamine does not block glucose-induced nuclear accumulation of MondoA:Mlx complexes or their binding to the TXNIP promoter. Rather, glutamine triggers the recruitment or activation of a histone deacetylase (HDAC) complex(s) to the TXNIP promoter, driving transcriptional repression. MondoA:Mlx complexes are strong activators of TXNIP gene expression; therefore, they may recruit HDAC complexes in the presence of glutamine. However, the involvement of other transcription factor(s) in HDAC corepressor recruitment cannot be ruled out at this time. If so, glutamine-targeted repression must be dominant over MondoA-dependent activation of TXNIP.

Glutamine has several intracellular fates; thus, there are several potential mechanisms by which it could drive TXNIP gene repression. Several experiments point to an important role for α-KG, which can be generated from glutamine [63]. For example, aminooxyacetate (AOA), which inhibits the transaminases required for the production of α-KG, blocks glutamine-dependent repression of TXNIP. Importantly, cell permeable analogs of α-KG restore α-KG levels in glutamine-depleted or AOA-treated cells and completely phenocopy glutamine-dependent repression of TXNIP. Collectively, these findings implicate α-KG as a key mediator of glutamine-dependent repression of TXNIP.

How does α-KG drive repression of TXNIP gene expression? There are several possibilities. First, α-KG can replenish the TCA cycle, raising the possibility that TXNIP promoter activity is directly linked to this central metabolic process. In this regard, the transient association of MondoA:Mlx with OMM places them in the perfect cellular locale to sense mitochondrial cues [128]. A second possibility is that α-KG is a cofactor for a number of dioxygenases that have established or emerging roles in controlling gene expression or epigenetics; thus, a-KG may repress TXNIP expression by a more indirect mechanism [102]. Determining the site of action for α-KG-dependent repression of TXNIP and whether α-KG or a α-KG -derived metabolite is sufficient for TXNIP repression will help reveal the mechanistic details.

Myc and MondoA Coordinate Nutrient Availability and Utilization

The Myc proto-oncogene has pleiotropic function in controlling glycolysis, yet its broad reach in controlling cell metabolism extends to mitochondrial activity and glutaminolysis [20, 29]. Myc controls glutaminolysis by upregulating glutamine transporters and glutaminase, suggesting that Myc-overexpressing cells can preferentially utilize glutamine as a carbon source to fuel biosynthetic reactions [43, 162]. Given the many cellular functions of glutamine, it is not surprising that Myc overexpression renders several cell types sensitive to glutamine depletion or inhibition of glutaminase [75, 82, 120, 165]. At this time, it is less clear which function of glutamine is most important for its role in promoting growth. It seems most likely that the essential functions of glutamine are different in different cell types.

Myc overexpression drives glutamine utilization, and glutaminolysis represses TXNIP expression [63]; thus, it is possible that the high rate of glutaminolysis in Myc-overexpressing cells might enhance glutamine-dependent repression of TXNIP (Fig. 5.2). Low TXNIP expression would facilitate aerobic glycolysis to further support cell growth. The coordinated regulation of TXNIP by Myc and MondoA represents a novel mechanism to control nutrient availability and utilization. Further, indirect repression of TXNIP by Myc represents a novel mechanism by which Myc could drive aerobic glycolysis independent of it's better-known function in driving expression of glycolytic target genes. Glutamine-dependent repression of TXNIP is not observed in all cell types (our unpublished data); thus, it will be of interest to determine whether it correlates exclusively with Myc pathway status. Triple negative breast cancers have high Myc levels, high rates of aerobic glycolysis, a glycolytic gene signature, and express glutaminase [13, 37, 56, 83, 104]. Further, triple negative breast cancer cell lines are more dependent on glutamine for growth than luminal cancer cell lines [72]. No targeted therapies are currently available for triple negative breast cancers; however, these finding suggest that glutaminase inhibitors such as BPTES, or BPTES derivatives, may have some utility [75].

Other Stresses that Regulate TXNIP

Many different signals induce TXNIP transcription, e.g., dexamethasone, vitamin D3, ER stress, hyperglycemia, heat shock, ROS, cell density, UV, serum removal, and LA (Fig. 5.2b). Thus, TXNIP is a broad sensor of cellular stress and likely restricts growth until homeostasis can be restored. It is also likely that proapoptotic activity of TXNIP contributes to cell death when a given stress exceeds its threshold. In a few instances, the transcriptional effector required to induce TXNIP in response to a specific stimulus has been identified. For example, an intact glucocorticoid response element (GRE), located 850 base pairs upstream of the TXNIP transcriptional start site, is required for its induction by dexamethasone [156]. Luciferase-based reporter assays suggest this GRE does not account for full induction observed for

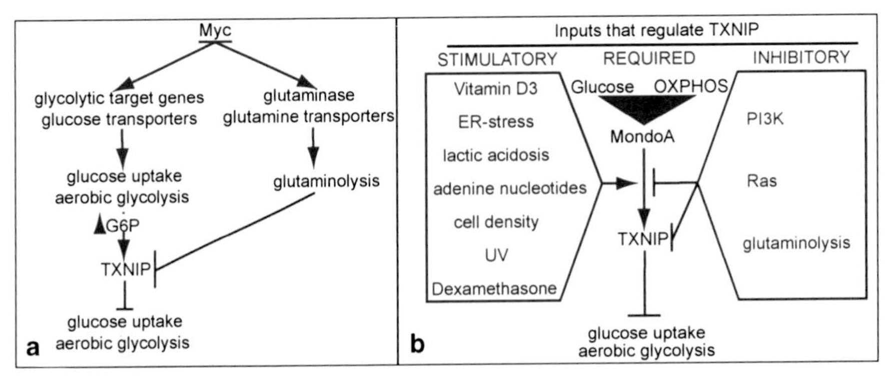

Fig. 5.2 The MondoA–TXNIP axis integrates oncogenic and metabolic signals. **a** Myc overexpression can drive both glycolysis and glutaminolysis. Our work shows that in some cell types, TXNIP expression is repressed by glutamine via a mechanism that requires its conversion to α-KG. This suggests that in cells with high levels of Myc-driven glutaminolysis, TXNIP levels will be repressed. Lower TXNIP levels are permissive for glucose uptake and aerobic glycolysis, providing another mechanism by which Myc dysregulation can drive aerobic glycolysis. **b** The MondoA–TXNIP axis functions as a negative feedback mechanism to restrict glucose uptake. Activation of the axis requires central bioenergetic signals, e.g., G6P and OXPHOS. TNXIP levels are elevated by a number of stimulatory stresses. The strict dependence of TXNIP expression on MondoA suggests that MondoA plays a role in the cellular stress response. However, other models are possible (see text). The output of the MondoA–TXNIP axis is inhibited by a number of common oncogenic signals, suggesting one mechanism by which cancer cells increase aerobic glycolysis

endogenous *TXNIP*, implying that other transcriptional regulators also participate in the dexamethasone response [156]. Our published and unpublished work shows that MondoA:Mlx complexes and glucose are strictly required for TXNIP induction in response to a spectrum of stress inducers, including LA [17, 63, 115, 145]. This finding suggests two models: first, different stresses may signal directly to MondoA:Mlx complexes through a common mechanism to activate TXNIP expression or, second, different stresses may signal through specific transcription factors, yet these factors must also require MondoA:Mlx complexes as permissivity factors to drive full transcriptional induction of TXNIP. Regardless of the precise model, our findings suggest that MondoA:Mlx complexes play a critical role in the cellular response to a variety of stresses. In extreme cases of no glucose, e.g., far from the blood supply, or no MondoA, e.g., inactivating mutation, we suggest that cells are incapable of activating TXNIP expression as a stress response (also see below).

Is the MondoA–TXNIP Axis Tumor/Growth Suppressive?

MondoA or TXNIP loss is sufficient to reprogram metabolism toward aerobic glycolysis. Thus, it is not surprising that expression studies implicate TXNIP as growth/tumor suppressor in a number of different cancers, including hepatocellular

MondoA downregulated

Cancer	Fold Change	P-value
Superficial Bladder Cancer (Sanchez-Carbayo)	-4.53	6.65E-12
Infiltrating Bladder Urothelial Carcinoma (Sanchez-Carbayo)	-2.37	3.70E-10
Prostatic Intraepithelial Neoplasia (Tomlins)	-2.70	1.53E-6
Burkitt's Lymphoma (Basso)	-2.50	1.45E-5
Rectal Mucinous Carcinoma (TCGA)	-2.22	4.43E-5
Colon Mucinous Carcinoma (TCGA)	-2.01	1.10E-8

MondoA upregulated

Cancer	Fold Change	P-value
B-Cell Acute Lymphoblastic Leukemia (Haferlach)	3.46	1.32E-58
B-Cell Childhood Acute Lymphoblastic Leukemia (Haferlach)	3.43	1.29E-69
Prostate Adenocarcinoma (Wallace)	3.63	7.17E-5
Clear Cell Renal Cell Carcinoma (Jones)	3.08	4.86E-15

Mlx downregulated

Cancer	Fold Change	P-value
Invasive Breast Carcinoma (Finak)	-7.05	4.29E-19
Colon Adenocarcinoma (TCGA)	-2.07	6.02E-15
Colon Mucinous Adenocarcinoma (TCGA)	-2.00	4.16E-13
Rectal Adenocarcinoma (TCGA)	-2.05	9.06E-15
Rectal Mucinous Adenocarcinoma (TCGA)	-2.77	7.43E-5
Acute Myeloid Leukemia (Stegmaier)	-3.48	4.09E-5

Fig. 5.3 MondoA and Mlx expression in tumor datasets. Oncomine (www.oncomine.org) was used to examine changes in MondoA and Mlx mRNA expression in different cancers. Upregulation or downregulation is listed along with the *P* value. Following each cancer type is the first author of the primary research study. The publication is listed in the references. (TCGA; The Cancer Genome Atlas)

carcinoma, breast cancer, bladder cancer, gastric cancer, and leukemia [167]. The putative role of TXNIP as a tumor suppressor has been best studied in breast cancer. For example, in one study of 788 patients [12], high TXNIP mRNA expression portended a better prognosis. A second study examined four large datasets with 885 patients also showed that high TXNIP and high ARRDC4 expression correlated with increased survival.

Because, TXNIP expression is strictly dependent on MondoA, we have suggested that MondoA may also function as a growth/tumor suppressor. Supporting this hypothesis, MondoA mRNA levels are suppressed in several types of lymphoma, including Burkitt's lymphoma that typically has dysregulated Myc expression, superficial bladder cancer, infiltrating bladder urothelial carcinoma, colorectal carcinoma, rectal and colon mucinous adenocarcinoma, prostatic intraepithelial neoplasia, and several types of sarcoma [5, 127, 147] (The Cancer Genome Atlas; Fig. 5.3). TXNIP mRNA levels are also generally lower in many of these same tumors (data not shown), suggesting that the MondoA–TXNIP axis may be impaired in these cancer types. Given the protective role of TXNIP in breast cancer, it is somewhat surprising that MondoA mRNA levels are not also generally suppressed in breast cancer. However, some breast tumors have very low MondoA levels, suggesting a subclass-specific loss of function [9]. Further, it is probable that negative regulation of MondoA nuclear activity by oncogenic signaling pathways (see below) is more relevant to its putative growth-suppressive activity than its mRNA expression level. It is intriguing that Mlx, which is an obligate partner for MondoA [115], mRNA levels are reduced in breast cancer [36] (The Cancer Genome Atlas; Fig. 5.3), suggesting another route to inactivation loss of the MondoA–TXNIP circuit.

In contrast to the simple model that MondoA is a growth/tumor suppressor, it is upregulated in some cancer types, including B cell acute lymphoblastic leukemia [51] (B-ALL; Fig. 5.3). MondoA function in this context appears to be quite different in that it promotes glucose uptake in B-ALL cell lines rather than suppresses glucose uptake [160]. MondoA expression also correlated with a less differentiated cell fate and regulated a "stemness" gene expression program that is anti-correlated with a good prognosis signature for B-ALL patients treated with glucocorticoids. This finding contrasts other reports, indicating that TXNIP contributes to glucocorticoid-driven apoptosis in a number of T cell leukemia cell lines [156]. Perhaps, MondoA can't regulate TXNIP in B-ALL, and this tumor-/growth-suppressive axis is not intact.

Oncogenic Signaling Pathways Abrogate the MondoA–TXNIP Axis

When stimulated, quiescent cells reprogram their metabolism to support the biosynthetic reactions required to transit the G1 phase of the cell cycle and enter S phase [105, 157]. Thus, the metabolism in early- to mid-G1 resembles that of transformed cancer cells. TXNIP is strongly upregulated in G0, yet its protein level is downregulated within minutes following serum addition and G0 exit [31]. This downregulation parallels a rapid increase in glucose uptake and glycolysis indicative of metabolic reprogramming during early G1. Importantly, constitutive ectopic expression of MondoA or TXNIP slows progression into S phase, indicating that TXNIP downregulation not only correlates with this transition, but also is necessary for cells to transition G1. Serum addition restricts TXNIP transcription and translation simultaneously via signaling pathways that are frequently activated in cancer; PI3K signaling blocks MondoA-dependent transcriptional activation of TXNIP, whereas Ras signaling blocks translation of preexisting TXNIP mRNA (Fig. 5.2b) [31]. Collectively then, abrogation of the MondoA–TXNIP axis is a required component of the metabolic reprogramming toward aerobic glycolysis, which is required to transition from quiescence to S phase.

In our studies, activation of PI3K and Ras contributes to the downregulation of the MondoA–TXNIP circuit (Fig. 5.2b). Given the preponderance of PI3K and Ras activation in cancer and their established function in driving aerobic glycolysis [133, 136], we propose that their suppression of the MondoA–TXNIP circuit may be a fairly general feature of the metabolic phenotype of cancer. From a cancer perspective, we suggest that the MondoA–TXNIP regulatory circuit constitutes a metabolic or stress checkpoint that restricts cell growth. In non-transformed cells, we propose that in response to a variety of stresses MondoA-dependent upregulation of TXNIP restricts cell growth until the stress can be resolved. By contrast, cells with constitutive activation of the PI3K pathway and/or Ras, TXNIP levels remain low, allowing these cells to bypass the metabolic/stress checkpoint.

Gene Expression Changes in Response to the Tumor Microenvironment: LA

The LA tumor microenvironment clearly affects several different aspects of tumorigenesis, yet the key effectors of LA in promoting tumorigenesis and tumor progression remain largely unknown. An important step in filling this knowledge gap was the recent report describing how cells adapt at the transcriptional level to LA [16]. This study, which was conducted in human mammary epithelial cells (HMECs), revealed that LA (25 mM lactic acid, pHe 6.7) upregulated gene sets representing nutrient deprivation and good breast cancer prognosis, and depleted gene sets representing E2F1 targets, mitotic cyclins, and poor breast cancer prognosis. The authors established a LA gene signature and found that, in contrast to the general perception of LA as a positive regulator tumorigenesis, a strong LA gene signature associated with favorable clinical outcomes in breast cancer patients. Further, the LA gene signature was higher in tumors with wild-type p53 compared to tumors with mutant p53 and high LA pathway activity correlated with less aggressive growth in xenograft studies. Thus, in HMECs, LA drives a protective gene expression program that is p53 dependent.

What are the key downstream pathways that mediate the protective effects of LA? Examination of tumor gene expression data revealed that tumors with the highest LA gene expression signature were enriched for pathways for biological processes such as TCA cycle, OXPHOS, and metabolism of fatty acids and amino acids. Further, LA repressed expression of several glycolytic enzymes. Thus, LA in HMECs drives metabolic reprogramming by repressing a glycolytic gene expression program and favoring a gene expression program that features aerobic respiration. Further, the LA gene signature is similar to those derived from cells treated with PI3K inhibitors, which fits well with the established function of PI3K in driving glycolysis. These findings are consonant with others, showing that inhibition of glycolysis and shifting metabolism toward aerobic respiration can restrict proliferation and tumorigenesis. Such a mechanism may explain why women with tumors with a high LA expression signature have a more favorable prognosis than those that don't.

LA drives a nutrient depletion gene expression signature, so it is not surprising that LA treatment activates AMPK, a sensor of low nutrient status, and inactivates mammalian target of rapamycin complex 1 (mTORC1) [16]. AMPK activation typically stimulates glucose uptake [17], so it is paradoxical that LA represses glucose uptake. To help resolve this paradox, the Chi group identified a set of 115 genes that were reciprocally regulated by LA and glucose deprivation. Importantly, similar to the full LA-dependent gene signature, this smaller gene signature was also predictive of good outcome in breast cancer [17]. TXNIP and ARRDC4 were among the most differentially expressed genes, being upregulated by LA and downregulated in the absence of glucose. This later finding fits well with our work, showing that the expression of TXNIP and ARRDC4 is strictly dependent on glucose. The LA-dependent induction of TNXIP and ARRDC4 was strongly, if not entirely,

MondoA dependent. Further, the LA-dependent restriction of glucose uptake required MondoA and TXNIP. Reduction of MondoA or TXNIP did not fully counter the LA-dependent reduction of glucose uptake, implicating a requirement for other pathways. In this study, acidosis was the primary driver of TXNIP expression, and this activity was further stimulated by lactate. LA increased the level of MondoA at the carbohydrate response elements in the proximal region of the TXNIP promoter, suggesting the mechanism by which LA increases TXNIP expression. Additional experiments are necessary to determine whether LA increases the nuclear accumulation of MondoA, its promoter binding, or its function as a transcriptional activator. Collectively, these studies suggest the LA-dependent increase of MondoA at the TXNIP promoter increases TXNIP expression, which contributes to the observed reduction in glucose uptake, although other pathways also contribute to this blockade. Underlying the importance of the LA-dependent activation of TXNIP and ARRDC4 is the finding that their expression correlates well with the broader LA-dependent gene signature and their high expression correlates with better prognosis. Finally, we suggest a tumor-/growth-suppressive role of MondoA–TXNIP axis in mediating a reprogramming of metabolism away from aerobic glycolysis under LA.

Potential Mechanisms by Which LA Regulates MondoA Transcriptional Activity

LA upregulates TXNIP transcription by increasing MondoA occupancy at the TXNIP promoter. In this study, low pHe drives TXNIP expression, and this activity was further stimulated by lactate. LA may stimulate nuclear accumulation or promoter binding of MondoA at the TXNIP promoter [17]. However, glucose also increases transactivation function of MondoA [115]; thus, LA may also stimulate the recruitment of coactivators. In this section, we speculate on potential models by which LA may increase MondoA activity at the TXNIP promoter. Further studies are necessary to test these different models.

Proton Sensing by MondoA

MondoA may respond to pH by sensing protons directly or indirectly. For example, deletion of MCRI, which contains three highly conserved histidines, blocks nuclear retention of MondoA in response to glucose [115]. Histidine is the only H^+ titratable residue within the physiological pH range and has been implicated as a pH sensor in many proteins [4, 84, 122, 155, 158]; thus, it is possible that the MCRI histidines regulate the transcriptional activity of MondoA. Supporting this model, mutation of the MCRI histidines (H78, H81, and H88) to alanine renders MondoA constitutively cytoplasmic and transcriptionally inactive even in the face of elevated glucose (data not shown). Furthermore, protein–lipid and protein–protein interactions can be regulated by pH. For example, the pleckstrin homology (PH) domain of Dbs has lower affinity for membrane phospholipid $PI_{(4,5)}P_2$ at pHi > 7.2 [38], and the focal adhesion protein talin has decreased binding to actin filaments at pHi > 7.2 [141]. MondoA lacks known phosphoinositide-interacting domains such as PH domains and FYVE; however, a fraction of MondoA is membrane associated (our unpublished data) [128]. This raises the possibility that low pHi might affect MondoA's association with intracellular membranes or membrane proteins and control its function more indirectly. One candidate for a membrane-associated MondoA-binding protein is v-ATPase. v-ATPase localizes to the lysosomal outer membrane and couples intracellular and organellar pH to different cellular processes by conformation-specific interactions with specific proteins or complexes, e.g., mTORC1 [125, 169]. It is intriguing that in yeast, with some support in mammalian cells, that the assembly of v-ATPase is glucose dependent with assembly occurring in a time frame similar to MondoA's glucose-dependent nuclear accumulation and promoter binding [67, 130]. Thus, it is possible that v-ATPase couples the lysosomal pH or pHi with the nuclear accumulation and transcriptional activity of MondoA.

Low pHe Drives Histone Hypoacetylation

A recent report demonstrates that lowering pHe leads to hypoacetylation of the amino-terminal tails of histones, providing a coupling between the extracellular environment, chromatin, and gene expression [91]. While histone hypoacetylation is generally considered a gene repression mechanism, the effects of bulk hypoacetylation on the expression of specific targets are generally unknown. It is worth testing whether alterations of the epigenetic landscape of the TXNIP promoter following LA contribute to its upregulation.

Low pH Increases Ca^{2+}_i

pH can modulate intracellular Ca^{2+} pools by several mechanisms. For example, low pHi stimulates ER-derived Ca^{2+} release by affecting the Ca^{2+} channel activity [26, 27]. Further, low pHi in cardiac myocytes following ischemia leads to activation of NHE to restore the pHi resulting in intracellular Na^+ overload. This leads to intracellular Ca^{2+} overload by activating the Na^+/Ca^{2+} antiporter [52]. Studies in pancreatic β-cells showed that Ca^{2+} can regulate TXNIP by activating sorcin, which drives ChREBP nuclear accumulation [76, 101]. Furthermore, in HeLa cells, several adenine-containing nucleotides trigger MondoA activation of TXNIP using a mechanism that required Ca^{2+} release [164]. Thus, low pHe can change intracellular Ca^{2+} levels by one or multiple mechanism, which may contribute to the LA-dependent increase in TXNIP and ARRDC4.

PI3K/Akt Signaling

Activation of PI3K signaling blocks MondoA-dependent transcriptional activation of TXNIP by blocking binding of MondoA to the TXNIP promoter [31]. Related to this is the finding that the gene signature induced by LA, which includes TXNIP and ARRDC4, is similar to gene expression signatures derived following treatment of PI3K inhibitors [16]. Together, these two findings suggest that by unspecified mechanisms, LA may result in a blockade of PI3K signaling which would allow for elevated TXNIP expression.

ER Stress

Acidosis induces ER stress in different cell lines, e.g., mouse astrocytes, mesothelial cells, and human glioblastoma cell lines [3, 61]. Recently, multiple groups showed that TXNIP expression is induced by ER stress through the PERK and IRE1 pathways in pancreatic β-cells [2, 77, 103]. Perhaps, pHe-induced ER stress mediates the LA-induced TXNIP expression.

G Protein-Coupled Receptor Signaling

A group of structurally related G protein-coupled receptors (GPCR; e.g., TDAG8, OGR1, and GPR4) function as H^+ sensors using multiple histidines in their extracellular domains, and mediate the downstream signaling by activating the adenylyl cyclase/cAMP/PKA cascade [58, 84, 122, 155]. Although controversial, one group has reported the transcriptional activation of ChREBP involves PKA-mediated phosphorylation [65, 68, 148]. The proposed PKA phosphorylation sites in ChREBP are not conserved in MondoA. Nonetheless, given that GPCRs can sense and respond to decrease in pHe, it is worthwhile testing whether signals downstream of GPCR signaling mediate MondoA-dependent activation of TXNIP.

mRNA Stabilization

mRNA encoding PEPCK, glutaminase, and glutamate dehydrogenase are selectively stabilized upon acidosis. This low pH-responsive stabilization is conferred by differential RNA binding of coregulator proteins to the conserved adenylate uridylate (AU)-rich sequences in the 3′ untranslated region (3′UTR) of these messages. This results in a decreased recruitment of deadenylase to the polyA tail of mRNA, which prevents mRNA from degradation upon deadenylation [50, 57, 74, 99]. Although no AU-rich elements were found within the +692–+1204 region of human TXNIP 3′UTR [168], there are AU-rich sequences present more upstream in the 3′UTR human TXNIP mRNA that could function as pH-response elements (our unpublished analysis). Thus, it would be interesting to examine whether similar mechanisms result in TXNIP mRNA stabilization under LA conditions, providing another level of regulation to maximize TXNIP expression in response to acidosis.

Does the MondoA–TXNIP Axis Function as a Bioenergetic Checkpoint?

Checkpoints allow cells to respond to a variety of insults by arresting division and allowing damage, repair, or restoration of homeostatic function. First observed in control of the cell cycle, it is now clear that checkpoint pathways are also in place to allow cells to respond to an energy-depleted state. For example, liver kinase B1 (LKB1) and AMPK sense high AMP levels and trigger an orchestrated response where catabolic pathways are activated and anabolic pathways are repressed to restore bioenergetic homeostasis [53]. Similarly, general control nonderepressible 2 (GCN2) senses uncharged transfer RNA (tRNA) molecules as a proxy for low amino acid levels, blocking translation and stimulating the synthesis of amino acids [42]. Rather than sensing an energy-depleted state, our data suggest that MondoA–TXNIP axis constitutes a checkpoint response that is activated by microenvironment stresses like LA and an energy-replete state such as high glycolysis. Activation of this checkpoint restricts glucose uptake, aerobic glycolysis, and cell growth in order to resolve the microenvironmental stress and normalize intracellular metabolism. Like other checkpoints in cancer, we suggest the MondoA–TXNIP axis is intact in cells early in tumor progression resulting in high TXNIP, which portends a better outcome for patients. Likewise, we suggest the MondoA–TXNIP axis can't be fully activated in tumors later in tumor progression resulting in low TXNIP, which portends a poor outcome for patients. As discussed above, the transcriptional function of MondoA and translation of TXNIP mRNA are under the negative control of signal pathways that are frequently activated in human cancer, providing a route to inactivate the MondoA–TXNIP axis.

Metabolic Checkpoints and Control of the Lactic Acidotic Microenvironment: Working Models

How the metabolic state of tumor cells affects its stromal neighbors and vice versa is just coming into focus and is an active area of research. As above, several possibilities exist for metabolic cooperation between tumor and stroma, including metabolic coupling and the reverse Warburg effect. Our data on the MondoA–TXNIP checkpoint suggest it may have cell-autonomous and cell-nonautonomous roles in dictating the metabolic "state" of the tumor microenvironment. Further, given that

MondoA–TXNIP can respond to LA, it is also likely that it allows cells to respond to changes in the microenvironment. We outline some working models for the MondoA–TXNIP checkpoint in different cellular contexts (Fig. 5.4).

From a cell-autonomous standpoint, we propose that an intact MondoA–TXNIP axis is one component of the cellular response to elevated glucose uptake or glycolysis. We envision that in epithelial cells with glycolysis-driving oncogenic events or that are far from a blood supply, that activation of the MondoA–TXNIP checkpoint restricts glucose uptake and restricts aerobic glycolysis. This restriction of glycolysis would limit the availability of glucose-derived carbons required for growth permissive biosynthesis. Further, restriction of glycolysis would reduce secretion of lactate, resulting in a normalization of LA in the microenvironment. This would disrupt metabolic coupling between tumor and stroma cells and abrogate the pro-growth/invasion phenotypes associated with the low pH microenvironment. In addition, restriction of LA in the microenvironment should prevent the metabolic reprogramming of tumor-associated fibroblasts in the case of the reverse Warburg model. Instead, we suggest that stromal fibroblasts also respond to LA by upregulating TXNIP, restricting rather than activating aerobic glycolysis. In cases of severe LA where homeostasis cannot be restored, TXNIP may contribute to apoptosis in the stromal cells.

While not yet elucidated with mechanistic detail, our work, and that of others, suggests that many pathways associated with cell transformation negatively regulate the MondoA–TXNIP checkpoint. For example, PI3K pathway activation, Ras activation, and elevated glutaminolysis are very common in cancer, and each acts as repressors of MondoA activity and/or TXNIP expression (Fig. 5.2b). Thus, we suggest that the MondoA–TXNIP checkpoint is inactivated or attenuated during tumor progression, blocking the cellular response to a variety of different stresses, including glycolytic flux and LA. Inactivation of the MondoA–TXNIP checkpoint has several implications for tumor cells and the surrounding environment. The inability of the tumor cell to attenuate aerobic glycolysis in the face of mounting LA would support unchecked biosynthesis and elevated secretion of lactate: a positive feed-forward regulatory circuit. Increasing LA would lower the pH of the microenvironment, increase the invasive potential of the tumor cells, and allow for the clonal evolution of surrounding cells that acquired resistance to apoptosis. Further, increased lactate would increase metabolic coupling with surrounding stromal cells or increase metabolic reprogramming of stromal fibroblasts in the case of the reverse Warburg model. From a different perspective, it is also possible that highly metabolic tumor cells with elevated glycolysis and glutaminolysis may restrict the growth of their non-transformed neighbors by utilizing the bulk of available nutrients [97]. Such a model predicts the clonal evolution of cells harboring mutations that are insensitive to nutrient limitation. Cells with mutations in LKB1 or TSC2 fail to respond to nutrient deprivation, representing examples of this class of mutant [59, 134, 135].

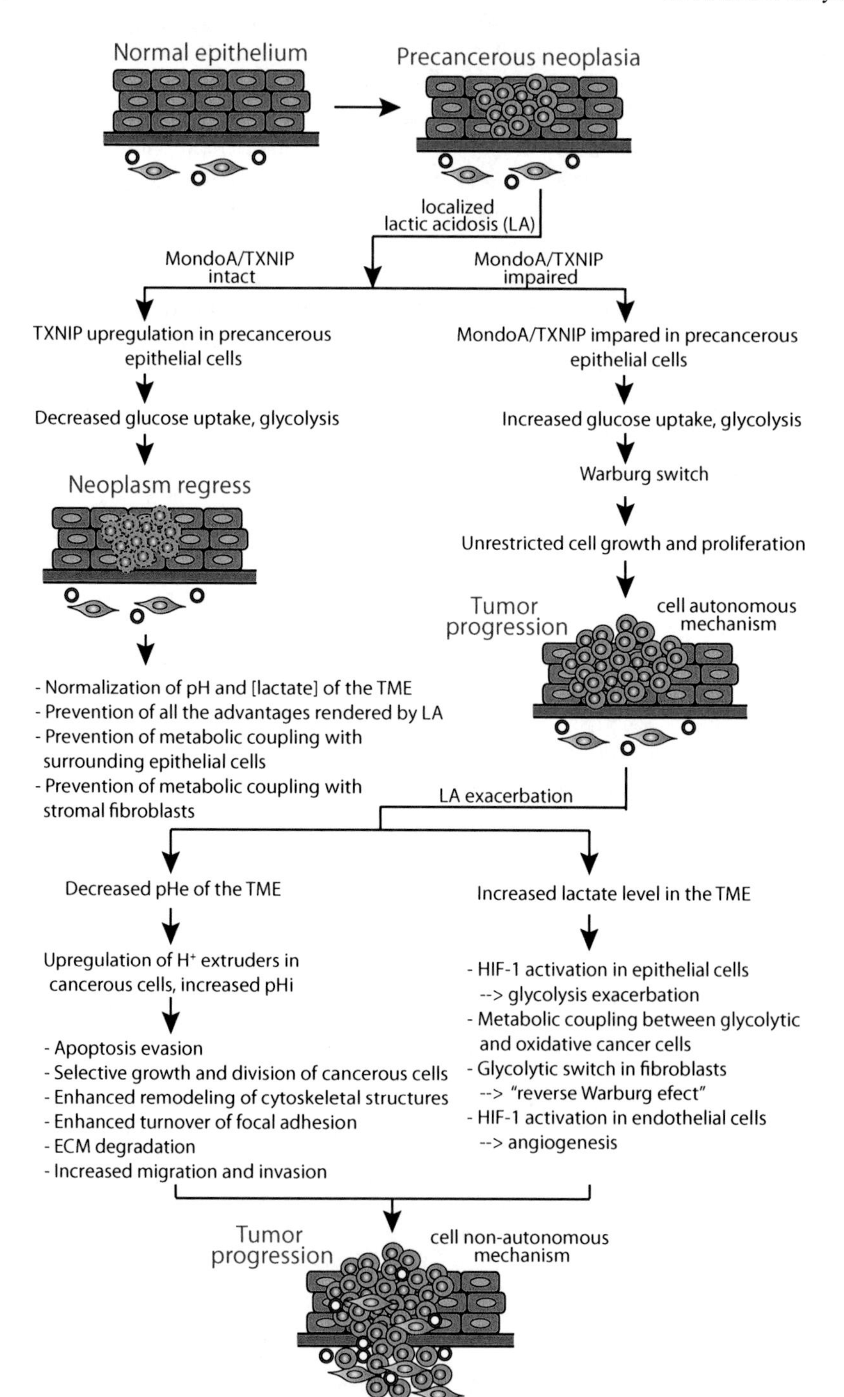

Fig. 5.4 The MondoA–TXNIP axis and the tumor microenvironment—working models. Our data suggest that the MondoA–TXNIP axis functions as a checkpoint that is activated by high glycolytic flux and extracellular stresses like LA. The checkpoint response, via restriction of glucose

Conclusions and Future Perspectives

How cells sense and respond transcriptionally to intracellular levels of different nutrients and their flux into and through different metabolic or biosynthetic pathways is critical for maintaining metabolic homeostasis. Disruption of these regulatory controls can contribute to metabolic reprogramming towards aerobic glycolysis, a common feature of the transformed state. The transcription factors Myc and HIF-1α, via their coordinated activation of most glycolytic and other metabolic target genes, are important drivers of aerobic glycolysis. By contrast, MondoA and its partner Mlx are just emerging as important sensors of glycolytic and glutaminolytic flux and they coordinate the use of glucose and glutamine. Unlike Myc and HIF-1α, MondoA:Mlx complexes, through their activation of TXNIP, restrict glucose uptake via a negative feedback mechanism. MondoA:Mlx complexes are also under the control LA, and this feedback mechanism would also normalize the pH of the microenvironment. Similar to other checkpoints, we propose an important role for MondoA in sensing and responding to intracellular and extracellular bioenergetic cues. When these cues fall outside of the physiological range, we suggest that MondoA triggers an adaptive transcriptional response that restricts cell growth until homeostasis can be restored. MondoA's regulation of TXNIP appears to play an important role in the cellular response to increased glycolytic flux and a number of stresses, including LA. A complete description of the direct MondoA-dependent transcriptome under different conditions will allow a deeper understanding of how cells coordinate their stress response.

If the MondoA–TXNIP checkpoint is truly growth/tumor suppressive, it will be important to determine whether it can be inactivated by oncogenic signals other than PI3K, Ras, and glutaminolysis. By doing so, it may be possible to identify novel therapeutic approaches that target cells based on the functional status of the checkpoint. Myc, HIF-1α, and TXNIP appear to be the central regulators of aerobic glycolysis in mammals, yet we only have a rudimentary understanding of how they coordinate glucose utilization. One simple model is that Myc and HIF-1α control glucose utilization by regulating glycolytic target genes and TXNIP controls access to glucose. Thus, it will also be important to determine the relative contribution of Myc and HIF-1α upregulation and TXNIP repression to aerobic glycolysis in the G0 to G1 transition and in cancer. We have focused this review primarily on TXNIP as a repressor of glucose uptake and aerobic glycolysis, yet it has other functions, e.g., HIF-1α destabilization, $p27^{Kip1}$ stabilization, inhibition of thioredoxin activity, apoptosis, and maintenance of mitochondrial function, which may also contribute to its growth-/tumor-suppressive activity. It will be important to elucidate the key

uptake and aerobic glycolysis, allows normalization of intracellular bioenergetic homeostasis and a reduction of LA in the tumor microenvironment. Common oncogenic lesions inactivate the MondoA–TXNIP checkpoint (Fig. 5.2b), allowing unrestricted glucose uptake and aerobic glycolysis, providing a cell-autonomous growth advantage. Unchecked aerobic glycolysis also exacerbates LA in the tumor microenvironment, which provides a number of cell-nonautonomous advantages as discussed in the text

TXNIP effectors in different cell contexts to fully appreciate its function in physiological and pathophysiological states.

Acknowledgments We thank members of the Ayer lab for their insights. R01GM055668-15, R01DK084425-04, and funds from the Huntsman Cancer Foundation support the work in our lab. The Cancer Center Support Grant P30 CA42014 supports core facility use at the Huntsman Cancer Institute.

References

1. Andres V, Carreras J, Cusso R (1990) Regulation of muscle phosphofructokinase by physiological concentrations of bisphosphorylated hexoses: effect of alkalinization. Biochem Biophys Res Commun 172:328–334
2. Anthony TG and Wek RC (2012) TXNIP switches tracks toward a terminal UPR. Cell Metab 16:135–137
3. Aoyama K, Burns DM, Suh SW, Garnier P, Matsumori Y, Shiina H, Swanson RA (2005) Acidosis causes endoplasmic reticulum stress and caspase-12-mediated astrocyte death. J Cereb Blood Flow Metab 25:358–370
4. Baird FE, Pinilla-Tenas JJ, Ogilvie WL, Ganapathy V, Hundal HS, Taylor PM (2006) Evidence for allosteric regulation of pH-sensitive System A (SNAT2) and System N (SNAT5) amino acid transporter activity involving a conserved histidine residue. Biochem J 397:369–375
5. Basso K, Margolin AA, Stolovitzky G, Klein U, Dalla-Favera R, Califano A (2005) Reverse engineering of regulatory networks in human B cells. Nat Genet 37:382–390
6. Billin AN, Ayer DE (2006) The Mlx network: evidence for a parallel Max-like transcriptional network that regulates energy metabolism. In: Eisenman RN (ed) The Myc/Max/Mad transcription factor network. Springer, Heidelberg, pp. 255–278.
7. Billin AN, Eilers AL, Queva C, Ayer DE (1999) Mlx, a novel max-like BHLHZip protein that interacts with the max network of transcription factors. J Biol Chem 274:36344–36350
8. Billin AN, Eilers AL, Coulter KL, Logan JS, Ayer DE (2000) MondoA, a novel basic helix-loop-helix-leucine zipper transcriptional activator that constitutes a positive branch of a max-like network. Mol Cell Biol 20:8845–8854
9. Boersma BJ, Reimers M, Yi M, Ludwig JA, Luke BT, Stephens RM, Yfantis HG, Lee DH, Weinstein JN, Ambs S (2008) A stromal gene signature associated with inflammatory breast cancer. Int J Cancer 122:1324–1332
10. Brisson L, Reshkin SJ, Gore J, Roger S (2012) pH regulators in invadosomal functioning: proton delivery for matrix tasting. Eur J Cell Biol 91:847–860
11. Brizel DM, Schroeder T, Scher RL, Walenta S, Clough RW, Dewhirst MW, Mueller-Klieser W (2001) Elevated tumor lactate concentrations predict for an increased risk of metastases in head-and-neck cancer. Int J Radiat Oncol Biol Phys 51:349–353
12. Cadenas C, Franckenstein D, Schmidt M, Gehrmann M, Hermes M, Geppert B, Schormann W, Maccoux LJ, Schug M, Schumann A et al (2010) Role of thioredoxin reductase 1 and thioredoxin interacting protein in prognosis of breast cancer. Breast Cancer Res 12, R44.
13. Cardone RA, Casavola V, Reshkin SJ (2005) The role of disturbed pH dynamics and the Na^+/H^+exchanger in metastasis. Nat Rev Cancer 5:786–795
14. Chandriani S, Frengen E, Cowling VH, Pendergrass SA, Perou CM, Whitfield ML, Cole MD (2009) A core MYC gene expression signature is prominent in basal-like breast cancer but only partially overlaps the core serum response. PLoS One 4:e6693
15. Chen J, Saxena G, Mungrue IN, Lusis AJ, Shalev A (2008a) Thioredoxin-interacting protein: a critical link between glucose toxicity and beta-cell apoptosis. Diabetes 57:938–944

16. Chen JL, Lucas JE, Schroeder T, Mori S, Wu J, Nevins J, Dewhirst M, West M, Chi JT (2008b) The genomic analysis of lactic acidosis and acidosis response in human cancers. PLoS Genet 4:e1000293
17. Chen JL, Merl D, Peterson CW, Wu J, Liu P, Yin H, Muoio DM, Ayer DE, West M, Chi J-T (2010) Lactic acidosis triggers starvation response with paradoxical induction of TXNIP through Mondo. A PLoS Genet 6:18
18. Chiche J, Le Fur Y, Vilmen C, Frassineti F, Daniel L, Halestrap AP, Cozzone PJ, Pouyssegur J, Lutz NW (2012) In vivo pH in metabolic-defective Ras-transformed fibroblast tumors: key role of the monocarboxylate transporter, MCT4, for inducing an alkaline intracellular pH. Int J Cancer 130:1511–1520
19. Chutkow WA, Patwari P, Yoshioka J, Lee RT (2008) Thioredoxin-interacting protein (Txnip) is a critical regulator of hepatic glucose production. J Biol Chem 283:2397–2406
20. Dang CV, Le A, Gao P (2009) MYC-induced cancer cell energy metabolism and therapeutic opportunities. Clin Cancer Res 15:6479–6483
21. De Saedeleer CJ, Copetti T, Porporato PE, Verrax J, Feron O, Sonveaux P (2012) Lactate activates HIF-1 in oxidative but not in Warburg-phenotype human tumor cells. PLoS One 7:e46571
22. DeBerardinis RJ, Cheng T (2010) Q's next: the diverse functions of glutamine in metabolism, cell biology and cancer. Oncogene 29:313–324
23. DeBerardinis RJ, Lum JJ, Hatzivassiliou G, Thompson CB (2008a) The biology of cancer: metabolic reprogramming fuels cell growth and proliferation. Cell Metab 7:11–20
24. Deberardinis RJ, Sayed N, Ditsworth D, Thompson CB (2008b) Brick by brick: metabolism and tumor cell growth. Curr Opin Genet Dev 18:54–61
25. Dentin R, Tomas-Cobos L, Foufelle F, Leopold J, Girard J, Postic C, Ferre P (2012) Glucose 6-phosphate, rather than xylulose 5-phosphate, is required for the activation of ChREBP in response to glucose in the liver. J Hepatol 56:199–209
26. Donoso P, Hidalgo C (1993) pH-sensitive calcium release in triads from frog skeletal muscle. Rapid filtration studies. J Biol Chem 268:25432–25438
27. Donoso P, Beltran M, Hidalgo C (1996) Luminal pH regulated calcium release kinetics in sarcoplasmic reticulum vesicles. Biochemistry 35:13419–13425
28. Egeblad M, Werb Z (2002) New functions for the matrix metalloproteinases in cancer progression. Nat Rev Cancer 2:161–174
29. Eilers M, Eisenman RN (2008) Myc's broad reach. Genes Dev 22:2755–2766
30. Eilers AL, Sundwall E, Lin M, Sullivan AA, Ayer DE (2002) A novel heterodimerization domain, CRM1, and 14–3-3 control subcellular localization of the MondoA-Mlx heterocomplex. Mol Cell Biol 22:8514–8526
31. Elgort MG, O'Shea JM, Jiang Y, Ayer DE (2010) Transcriptional and translational downregulation of Thioredoxin Interacting Protein is required for metabolic reprogramming during G1. Genes and Cancer 1:893–907
32. Eng CH, Yu K, Lucas J, White E, Abraham RT (2010) Ammonia derived from glutaminolysis is a diffusible regulator of autophagy. Sci Signal 3:ra31
33. Estrella V, Chen T, Lloyd M, Wojtkowiak J, Cornnell HH, Ibrahim-Hashim A, Bailey K, Balagurunathan Y, Rothberg JM, Sloane BF, et al (2013) Acidity generated by the tumor microenvironment drives local invasion. Cancer Res 73(5):1524-1535
34. Fais S, De Milito A, You H, Qin W (2007) Targeting vacuolar H^+-ATPases as a new strategy against cancer. Cancer Res 67:10627–10630
35. Fang JS, Gillies RD, Gatenby RA (2008) Adaptation to hypoxia and acidosis in carcinogenesis and tumor progression. Semin Cancer Biol 18:330–337
36. Finak G, Bertos N, Pepin F, Sadekova S, Souleimanova M, Zhao H, Chen H, Omeroglu G, Meterissian S, Omeroglu A et al (2008) Stromal gene expression predicts clinical outcome in breast cancer. Nat Med 14:518–527
37. Foulkes WD, Smith IE, Reis-Filho, JS (2010) Triple-negative breast cancer. N Engl J Med 363:1938–1948

38. Frantz C, Karydis A, Nalbant P, Hahn KM, Barber DL (2007) Positive feedback between Cdc42 activity and H^+ efflux by the Na-H exchanger NHE1 for polarity of migrating cells. J Cell Biol 179:403–410
39. Frantz C, Barreiro G, Dominguez L, Chen X, Eddy R, Condeelis J, Kelly MJ, Jacobson MP, Barber DL (2008) Cofilin is a pH sensor for actin free barbed end formation: role of phosphoinositide binding. J Cell Biol 183:865–879
40. Frieden C, Gilbert HR, Bock PE (1976) Phosphofructokinase. III. Correlation of the regulatory kinetic and molecular properties of the rabbit muscle enzyme. J Biol Chem 251:5644–5647
41. Furlong IJ, Ascaso R, Lopez Rivas A, Collins MK (1997) Intracellular acidification induces apoptosis by stimulating ICE-like protease activity. J Cell Sci 110(Pt 5):653–661
42. Gallinetti J, Harputlugil E, Mitchell JR (2013) Amino acid sensing in dietary-restriction-mediated longevity: roles of signal-transducing kinases GCN2 and TOR. Biochem J 449:1–10
43. Gao P, Tchernyshyov I, Chang TC, Lee YS, Kita K, Ochi T, Zeller KI, De Marzo AM, Van Eyk JE, Mendell JT et al (2009) c-Myc suppression of miR-23a/b enhances mitochondrial glutaminase expression and glutamine metabolism. Nature 458:762–765
44. Gatenby RA, Gillies RJ (2004) Why do cancers have high aerobic glycolysis? Nat Rev Cancer 4:891–899
45. Gatenby RA, Gillies RJ (2008) A microenvironmental model of carcinogenesis. Nat Rev Cancer 8:56–61
46. Gillies RJ, Raghunand N, Karczmar GS, Bhujwalla ZM (2002) MRI of the tumor microenvironment. J Magn Reson Imaging 16:430–450
47. Gordan JD, Simon MC (2007) Hypoxia-inducible factors: central regulators of the tumor phenotype. Curr Opin Genet Dev 17:71–77
48. Gordan JD, Thompson CB, Simon MC (2007) HIF and c-Myc: sibling rivals for control of cancer cell metabolism and proliferation. Cancer Cell 12:108–113
49. Gottlieb RA, Nordberg J, Skowronski E, Babior BM (1996) Apoptosis induced in Jurkat cells by several agents is preceded by intracellular acidification. Proc Natl Acad Sci U S A 93:654–658
50. Gummadi L, Taylor L, Curthoys NP (2012) Concurrent binding and modifications of AUF1 and HuR mediate the pH-responsive stabilization of phosphoenolpyruvate carboxykinase mRNA in kidney cells. Am J Physiol Renal Physiol 303:F1545–F1554
51. Haferlach T, Kohlmann A, Wieczorek L, Basso G, Kronnie GT, Bene MC, De Vos J, Hernandez JM, Hofmann WK, Mills KI, et al (2010) Clinical utility of microarray-based gene expression profiling in the diagnosis and subclassification of leukemia: report from the International Microarray Innovations in Leukemia Study Group. J Clin Oncol 28:2529–2537
52. Halestrap AP, Clarke SJ, Javadov SA (2004) Mitochondrial permeability transition pore opening during myocardial reperfusion–a target for cardioprotection. Cardiovasc Res 61:372–385
53. Hardie DG, Ross FA, Hawley SA (2012) AMPK: a nutrient and energy sensor that maintains energy homeostasis. Nat Rev Mol Cell Biol 13:251–262
54. Hashimoto T, Hussien R, Cho HS, Kaufer D, Brooks GA (2008) Evidence for the mitochondrial lactate oxidation complex in rat neurons: demonstration of an essential component of brain lactate shuttles. PLoS One 3:e2915
55. Havula E., Hietakangas V (2012) Glucose sensing by ChREBP/MondoA-Mlx transcription factors. Semin Cell Dev Biol 23:640–647
56. Horiuchi D, Kusdra L, Huskey NE, Chandriani S, Lenburg ME, Gonzalez-Angulo AM, Creasman KJ, Bazarov AV, Smyth JW, Davis SE et al (2012) MYC pathway activation in triple-negative breast cancer is synthetic lethal with CDK inhibition. J Exp Med 209:679–696
57. Ibrahim H, Lee YJ, Curthoys NP (2008) Renal response to metabolic acidosis: role of mRNA stabilization. Kidney Int 73, :11–18.
58. Ihara Y, Kihara Y, Hamano F, Yanagida K, Morishita Y, Kunita A, Yamori T, Fukayama M, Aburatani H, Shimizu T, et al (2010) The G protein-coupled receptor T-cell death-associated gene 8 (TDAG8) facilitates tumor development by serving as an extracellular pH sensor. Proc Natl Acad Sci USA 107:17309–17314

59. Inoki K, Zhu T, Guan KL (2003) TSC2 mediates cellular energy response to control cell growth and survival. Cell 115:577–590
60. Jeon JH, Lee KN, Hwang CY, Kwon KS, You KH, Choi I (2005) Tumor suppressor VDUP1 increases p27(kip1) stability by inhibiting JAB1. Cancer Res 65:4485–4489
61. Johno H, Ogata R, Nakajima S, Hiramatsu N, Kobayashi T, Hara H, Kitamura M (2012) Acidic stress-ER stress axis for blunted activation of NF-kappaB in mesothelial cells exposed to peritoneal dialysis fluid. Nephrol Dial Transplant 27:4053–4060
62. Jones RG, Plas DR, Kubek S, Buzzai M, Mu J, Xu Y, Birnbaum MJ, Thompson CB (2005) AMP-activated protein kinase induces a p53-dependent metabolic checkpoint. Molecular cell 18:283–293
63. Kaadige MR, Looper RE, Kamalanaadhan S, Ayer DE (2009) Glutamine-dependent anaplerosis dictates glucose uptake and cell growth by regulating MondoA transcriptional activity. Proc Natl Acad Sci U S A 106:14878–14883
64. Kaadige MR, Elgort MG, Ayer DE (2010) Coordination of glucose and glutamine utilization by an expanded Myc network. Transcription 1:36–40
65. Kabashima T, Kawaguchi T, Wadzinski BE, Uyeda K (2003) Xylulose 5-phosphate mediates glucose-induced lipogenesis by xylulose 5-phosphate-activated protein phosphatase in rat liver. Proc Natl Acad Sci U S A 100:5107–5112
66. Kalluri R, Zeisberg M (2006) Fibroblasts in cancer. Nat Rev Cancer 6:392–401
67. Kane PM (2012) Targeting reversible disassembly as a mechanism of controlling V-ATPase activity. Curr Protein Pept Sci 13:117–123
68. Kawaguchi T, Takenoshita M, Kabashima T, Uyeda K (2001) Glucose and cAMP regulate the L-type pyruvate kinase gene by phosphorylation/dephosphorylation of the carbohydrate response element binding protein. Proc Natl Acad Sci U S A 98:13710–13715
69. Kerkar SP, Restifo NP (2012) Cellular constituents of immune escape within the tumor microenvironment. Cancer Res 72:3125–3130
70. Kim JW, Tchernyshyov I, Semenza GL, Dang CV (2006) HIF-1-mediated expression of pyruvate dehydrogenase kinase: a metabolic switch required for cellular adaptation to hypoxia. Cell Metab 3:177–185
71. Ko YH, Lin Z, Flomenberg N, Pestell RG, Howell A, Sotgia F, Lisanti MP, Martinez-Outschoorn UE (2011) Glutamine fuels a vicious cycle of autophagy in the tumor stroma and oxidative mitochondrial metabolism in epithelial cancer cells: implications for preventing chemotherapy resistance. Cancer Biol Ther 12:1085–1097
72. Kung HN, Marks JR, Chi JT (2011) Glutamine synthetase is a genetic determinant of cell type-specific glutamine independence in breast epithelia. PLoS Genet 7:e1002229
73. Kuwata F, Suzuki N, Otsuka K, Taguchi M, Sasai Y, Wakino H, Ito M, Ebihara S, Suzuki K (1991) Enzymatic regulation of glycolysis and gluconeogenesis in rabbit periodontal ligament under various physiological pH conditions. J Nihon Univ Sch Dent 33:81–90
74. Laterza OF, Hansen WR, Taylor L, Curthoys NP (1997) Identification of an mRNA-binding protein and the specific elements that may mediate the pH-responsive induction of renal glutaminase mRNA. J Biol Chem 272:22481–22488
75. Le A, Lane AN, Hamaker M, Bose S, Gouw A, Barbi J, Tsukamoto T, Rojas CJ, Slusher BS, Zhang H et al (2012) Glucose-independent glutamine metabolism via TCA cycling for proliferation and survival in B cells. Cell Metab 15:110–121
76. Leclerc I, Rutter GA, Meur G, Noordeen N (2012) Roles of Ca^{2+} ions in the control of ChREBP nuclear translocation. J Endocrinol 213:115–122
77. Lerner AG, Upton JP, Praveen PV, Ghosh R, Nakagawa Y, Igbaria A, Shen S, Nguyen V, Backes BJ, Heiman M et al (2012) IRE1alpha induces thioredoxin-interacting protein to activate the NLRP3 inflammasome and promote programmed cell death under irremediable ER stress. Cell Metab 16:250–264
78. Li MV, Chang B, Imamura M, Poungvarin N, Chan L (2006) Glucose-dependent transcriptional regulation by an evolutionarily conserved glucose-sensing module. Diabetes 55:1179–1189

79. Li MV, Chen W, Harmancey RN, Nuotio-Antar AM, Imamura M, Saha P, Taegtmeyer H, Chan L (2010) Glucose-6-phosphate mediates activation of the carbohydrate responsive binding protein (ChREBP). Biochem Biophys Res Commun 395:395–400
80. Liao C, Hu B, Arno MJ, Panaretou B (2007) Genomic screening in vivo reveals the role played by vacuolar H^+ATPase and cytosolic acidification in sensitivity to DNA-damaging agents such as cisplatin. Mol Pharmacol 71:416–425
81. Liaudet-Coopman E, Beaujouin M, Derocq D, Garcia M, Glondu-Lassis M, Laurent-Matha V, Prebois C, Rochefort H, Vignon F (2006) Cathepsin D: newly discovered functions of a long-standing aspartic protease in cancer and apoptosis. Cancer Lett 237:167–179
82. Liu W, Le A, Hancock C, Lane AN, Dang CV, Fan TW, Phang JM (2012) Reprogramming of proline and glutamine metabolism contributes to the proliferative and metabolic responses regulated by oncogenic transcription factor c-MYC. Proc Natl Acad Sci U S A 109:8983–8988
83. Lopez-Knowles E, O'Toole SA, McNeil CM, Millar EK, Qiu MR, Crea P, Daly RJ, Musgrove EA, Sutherland RL (2010) PI3K pathway activation in breast cancer is associated with the basal-like phenotype and cancer-specific mortality. Int J Cancer 126:1121–1131
84. Ludwig MG, Vanek M, Guerini D, Gasser JA, Jones CE, Junker U, Hofstetter H, Wolf RM, Seuwen K (2003) Proton-sensing G-protein-coupled receptors. Nature 425:93–98
85. Ma L, Robinson LN, Towle HC (2006) ChREBP*Mlx is the principal mediator of glucose-induced gene expression in the liver. J Biol Chem 281:28721–28730
86. Martinez-Outschoorn UE, Balliet RM, Rivadeneira DB, Chiavarina B, Pavlides S, Wang C, Whitaker-Menezes D, Daumer KM, Lin Z, Witkiewicz AK et al (2010a) Oxidative stress in cancer associated fibroblasts drives tumor-stroma co-evolution: A new paradigm for understanding tumor metabolism, the field effect and genomic instability in cancer cells. Cell Cycle 9:3256–3276
87. Martinez-Outschoorn UE, Pavlides S, Whitaker-Menezes D, Daumer KM, Milliman JN, Chiavarina B, Migneco G, Witkiewicz AK, Martinez-Cantarin MP, Flomenberg N et al (2010b) Tumor cells induce the cancer associated fibroblast phenotype via caveolin-1 degradation: implications for breast cancer and DCIS therapy with autophagy inhibitors. Cell Cycle 9:2423–2433
88. Martinez-Outschoorn UE, Trimmer C, Lin Z, Whitaker-Menezes D, Chiavarina B, Zhou J, Wang C, Pavlides S, Martinez-Cantarin MP, Capozza F et al (2010c) Autophagy in cancer associated fibroblasts promotes tumor cell survival: Role of hypoxia, HIF1 induction and NFkappaB activation in the tumor stromal microenvironment. Cell Cycle 9:3515–3533
89. Martinez-Outschoorn UE, Pavlides S, Howell A, Pestell RG, Tanowitz HB, Sotgia F, Lisanti MP (2011) Stromal-epithelial metabolic coupling in cancer: integrating autophagy and metabolism in the tumor microenvironment. Int J Biochem Cell Biol 43:1045–1051
90. Matsuyama S, Reed JC (2000) Mitochondria-dependent apoptosis and cellular pH regulation. Cell Death Differ 7:1155–1165
91. McBrian MA, Behbahan IS, Ferrari R, Su T, Huang TW, Li K, Hong CS, Christofk HR, Vogelauer M, Seligson DB et al (2013) Histone Acetylation Regulates Intracellular pH. Molecular cell 49:310–321
92. McFerrin LG, Atchley WR (2011) Evolution of the Max and Mlx networks in animals. Genome Biol Evol 3:915–937
93. McFerrin LG, Atchley WR (2012) A novel N-terminal domain may dictate the glucose response of Mondo proteins. PLoS One 7:e34803
94. Meroni G, Cairo S, Merla G, Messali S, Brent R, Ballabio A, Reymond A (2000) Mlx, a new Max-like bHLHZip family member: the center stage of a novel transcription factors regulatory pathway? Oncogene 19:3266–3277
95. Migneco G, Whitaker-Menezes D, Chiavarina B, Castello-Cros R, Pavlides S, Pestell RG, Fatatis A, Flomenberg N, Tsirigos A, Howell A et al (2010) Glycolytic cancer associated fibroblasts promote breast cancer tumor growth, without a measurable increase in angiogenesis: evidence for stromal-epithelial metabolic coupling. Cell Cycle 9:2412–2422

96. Mohamed MM, Sloane BF (2006) Cysteine cathepsins: multifunctional enzymes in cancer. Nat Rev Cancer 6:764–775
97. Moreno E, Basler K (2004) dMyc transforms cells into super-competitors. Cell 117:117–129
98. Moseley JB, Okada K, Balcer HI, Kovar DR, Pollard TD, Goode BL (2006) Twinfilin is an actin-filament-severing protein and promotes rapid turnover of actin structures in vivo. J Cell Sci 119:1547–1557
99. Mufti J, Hajarnis S, Shepardson K, Gummadi L, Taylor L, Curthoys NP (2011) Role of AUF1 and HuR in the pH-responsive stabilization of phosphoenolpyruvate carboxykinase mRNA in LLC-PK(1)-F(+) cells. Am J Physiol Renal Physiol 301:F1066–F1077
100. Muller PA, Vousden KH (2013) p53 mutations in cancer. Nat Cell Biol 15:2–8
101. Noordeen NA, Meur G, Rutter GA, Leclerc I (2012) Glucose-induced nuclear shuttling of ChREBP is mediated by sorcin and Ca(2+) ions in pancreatic beta-cells. Diabetes 61:574–585
102. Oermann EK, Wu J, Guan KL, Xiong Y (2012) Alterations of metbolic genes and metabolites in cancer. Semin Cell Dev Biol 23:370–380
103. Oslowski CM, Hara T, O'Sullivan-Murphy B, Kanekura K, Lu S, Hara M, Ishigaki S, Zhu LJ, Hayashi E, Hui ST et al (2012) Thioredoxin-interacting protein mediates ER stress-induced beta cell death through initiation of the inflammasome. Cell Metab 16:265–273
104. Palaskas N, Larson SM, Schultz N, Komisopoulou E, Wong J, Rohle D, Campos C, Yannuzzi N, Osborne JR, Linkov I et al (2011) 18F-Fluorodeoxy-glucose Positron Emission Tomography Marks MYC-Overexpressing Human Basal-Like Breast Cancers. Cancer Res 71:5164–5174
105. Pardee AB (1974) A restriction point for control of normal animal cell proliferation. Proc Natl Acad Sci U S A 71:1286–1290
106. Parikh H, Carlsson E, Chutkow WA, Johansson LE, Storgaard H, Poulsen P, Saxena R, Ladd C, Schulze PC, Mazzini MJ et al (2007) TXNIP regulates peripheral glucose metabolism in humans. PLoS Med 4:e158
107. Park HJ, Lyons JC, Ohtsubo T, Song CW (1999) Acidic environment causes apoptosis by increasing caspase activity. Br J Cancer 80:1892–1897
108. Park HJ, Lyons JC, Ohtsubo T, Song CW (2000) Cell cycle progression and apoptosis after irradiation in an acidic environment. Cell Death Differ 7:729–738
109. Patwari P, Chutkow WA, Cummings K, Verstraeten VL, Lammerding J, Schreiter ER, Lee RT (2009) Thioredoxin-independent regulation of metabolism by the alpha-arrestin proteins. J Biol Chem 284:24996–25003
110. Patwari P, Lee RT (2012) An expanded family of arrestins regulate metabolism. Trends Endocrinol Metab 23:216–222
111. Pavlides S, Whitaker-Menezes D, Castello-Cros R, Flomenberg N, Witkiewicz AK, Frank PG, Casimiro MC, Wang C, Fortina P, Addya S et al (2009) The reverse Warburg effect: aerobic glycolysis in cancer associated fibroblasts and the tumor stroma. Cell Cycle 8:3984–4001
112. Pavlides S, Tsirigos A, Migneco G, Whitaker-Menezes D, Chiavarina B, Flomenberg N, Frank PG, Casimiro MC, Wang C, Pestell RG et al (2010) The autophagic tumor stroma model of cancer: Role of oxidative stress and ketone production in fueling tumor cell metabolism. Cell Cycle 9:3485–3505
113. Perez-Sala D, Collado-Escobar D, Mollinedo F (1995) Intracellular alkalinization suppresses lovastatin-induced apoptosis in HL-60 cells through the inactivation of a pH-dependent endonuclease. J Biol Chem 270:6235–6242
114. Peterson CW, Stoltzman CA, Sighinolfi MP, Han KS, Ayer DE (2010) Glucose controls nuclear accumulation, promoter binding, and transcriptional activity of the MondoA-Mlx heterodimer. Mol Cell Biol 30:2887–2895
115. Peterson CW, Ayer DE (2011) An extended Myc network contributes to glucose homeostasis in cancer and diabetes. Front Biosci 16:2206–2223

116. Petrie JL, Al-Oanzi ZH, Arden C, Tudhope SJ, Mann J, Kieswich J, Yaqoob MM, Towle HC, Agius L (2013) Glucose induces protein targeting to glycogen in hepatocytes by fructose 2,6-bisphosphate-mediated recruitment of MondoA to the promoter. Mol Cell Biol 33:725–738
117. Postic C, Dentin R, Denechaud PD, Girard J (2007) ChREBP, a tanscriptional regulator of glucose and lipid metabolism. Annu Rev Nutr 27:179–192
118. Pouyssegur J, Franchi A, L'Allemain G, Paris S (1985) Cytoplasmic pH, a key determinant of growth factor-induced DNA synthesis in quiescent fibroblasts. FEBS Lett 190:115–119
119. Putney LK, Barber DL (2003) Na-H exchange-dependent increase in intracellular pH times G2/M entry and transition. J Biol Chem 278:44645–44649
120. Qing G, Li B, Vu A, Skuli N, Walton ZE, Liu X, Mayes PA, Wise DR, Thompson CB, Maris JM et al (2012) ATF4 regulates MYC-mediated neuroblastoma cell death upon glutamine deprivation. Cancer Cell 22:631–644
121. Quennet V, Yaromina A, Zips D, Rosner A, Walenta S, Baumann M, Mueller-Klieser W (2006) Tumor lactate content predicts for response to fractionated irradiation of human squamous cell carcinomas in nude mice. Radiother Oncol 81:130–135
122. Radu CG, Nijagal A, McLaughlin J, Wang L, Witte ON (2005) Differential proton sensitivity of related G protein-coupled receptors T cell death-associated gene 8 and G2A expressed in immune cells. Proc Natl Acad Sci U S A 102:1632–1637
123. Raghunand N, Gillies RJ (2000) pH and drug resistance in tumors. Drug Resist Updat 3:39–47
124. Raghunand N, Gatenby RA, Gillies RJ (2003) Microenvironmental and cellular consequences of altered blood flow in tumours. Br J Radiol 76(Spec N 1):S11–22.
125. Recchi C, Chavrier P (2006). V-ATPase: a potential pH sensor. Nat Cell Biol 8:107–109
126. Rich IN, Worthington-White D, Garden OA, Musk P (2000) Apoptosis of leukemic cells accompanies reduction in intracellular pH after targeted inhibition of the Na(+)/H(+) exchanger. Blood 95:1427–1434
127. Sanchez-Carbayo M, Socci ND, Lozano J, Saint F, Cordon-Cardo C (2006) Defining molecular profiles of poor outcome in patients with invasive bladder cancer using oligonucleotide microarrays. J Clin Oncol 24:778–789
128. Sans CL, Satterwhite DJ, Stoltzman CA, Breen KT, Ayer DE (2006) MondoA-Mlx heterodimers are candidate sensors of cellular energy status: mitochondrial localization and direct regulation of glycolysis. Mol Cell Biol 26:4863–4871
129. Sansone P and Bromberg J (2011) Environment, inflammation, and cancer. Curr Opin Genet Dev 21:80–85
130. Sautin YY, Lu M, Gaugler A, Zhang L, Gluck SL (2005) Phosphatidylinositol 3-kinase-mediated effects of glucose on vacuolar H+-ATPase assembly, translocation, and acidification of intracellular compartments in renal epithelial cells. Mol Cell Biol 25:575–589
131. Schroder K, Zhou R, Tschopp J (2010) The NLRP3 inflammasome: a sensor for metabolic danger? Science 327:296–300
132. Shanware NP, Mullen AR, DeBerardinis RJ, Abraham RT (2011) Glutamine: pleiotropic roles in tumor growth and stress resistance. J Mol Med (Berl) 89:229–236
133. Shanware NP, Bray K, Abraham RT (2013) The PI3K, metabolic, and autophagy networks: interactive partners in cellular health and disease. Annu Rev Pharmacol Toxicol 53:89–106
134. Shaw RJ, Bardeesy N, Manning BD, Lopez L, Kosmatka M, DePinho RA, Cantley LC (2004a) The LKB1 tumor suppressor negatively regulates mTOR signaling. Cancer Cell 6:91–99
135. Shaw RJ, Kosmatka M, Bardeesy N, Hurley RL, Witters LA, DePinho RA, Cantley LC (2004b) The tumor suppressor LKB1 kinase directly activates AMP-activated kinase and regulates apoptosis in response to energy stress. Proc Natl Acad Sci U S A 101:3329–3335
136. Shaw RJ, Cantley LC (2006) Ras, PI(3)K and mTOR signalling controls tumour cell growth. Nature 441:424–430

137. Shin D, Jeon JH, Jeong M, Suh HW, Kim S, Kim HC, Moon OS, Kim YS, Chung JW, Yoon SR et al (2008) VDUP1 mediates nuclear export of HIF1alpha via CRM1-dependent pathway. Biochim Biophys Acta 1783:838–848
138. Sloan EJ and Ayer DE (2010) Myc, Mondo and Metabolism. Genes and Cancer 1:587–596
139. Sonveaux P, Vegran F, Schroeder T, Wergin MC, Verrax J, Rabbani ZN, De Saedeleer CJ, Kennedy KM, Diepart C, Jordan BF et al (2008) Targeting lactate-fueled respiration selectively kills hypoxic tumor cells in mice. J Clin Invest 118:3930–3942
140. Sonveaux P, Copetti T, De Saedeleer CJ, Vegran F, Verrax J, Kennedy KM, Moon EJ, Dhup S, Danhier P, Frerart F et al (2012) Targeting the lactate transporter MCT1 in endothelial cells inhibits lactate-induced HIF-1 activation and tumor angiogenesis. PLoS One 7:e33418
141. Srivastava J, Barreiro G, Groscurth S, Gingras AR, Goult BT, Critchley DR, Kelly MJ, Jacobson MP, Barber DL (2008) Structural model and functional significance of pH-dependent talin-actin binding for focal adhesion remodeling. Proc Natl Acad Sci U S A 105:14436–14441
142. Stock C, Schwab A (2009) Protons make tumor cells move like clockwork. Pflugers Arch 458:981–992
143. Stock C, Gassner B, Hauck CR, Arnold H, Mally S, Eble JA, Dieterich P, Schwab A (2005) Migration of human melanoma cells depends on extracellular pH and Na^+/H^+ exchange. J Physiol 567:225–238
144. Stoltzman CA, Peterson CW, Breen KT, Muoio DM, Billin AN, Ayer DE (2008) Glucose sensing by MondoA:Mlx complexes: a role for hexokinases and direct regulation of thioredoxin-interacting protein expression. Proc Natl Acad Sci U S A 105:6912–6917
145. Stoltzman CA, Kaadige MR, Peterson CW, Ayer DE (2011) MondoA senses non-glucose sugars: regulation of thioredoxin-interacting protein (TXNIP) and the hexose transport curb. J Biol Chem 286:38027–38034
146. Swartz MA, Iida N, Roberts EW, Sangaletti S, Wong MH, Yull FE, Coussens LM, and DeClerck YA (2012) Tumor microenvironment complexity: emerging roles in cancer therapy. Cancer Res 72:2473–2480
147. Tomlins SA, Mehra R, Rhodes DR, Cao X, Wang L, Dhanasekaran SM, Kalyana-Sundaram S, Wei JT, Rubin MA, Pienta KJ et al (2007) Integrative molecular concept modeling of prostate cancer progression. Nat Genet 39:41–51
148. Tsatsos NG, Davies MN, O'Callaghan BL, Towle HC (2008) Identification and function of phosphorylation in the glucose-regulated transcription factor ChREBP. Biochem J 411:261–270
149. Uyeda K, Repa JJ (2006) Carbohydrate response element binding protein, ChREBP, a transcription factor coupling hepatic glucose utilization and lipid synthesis. Cell Metab 4:107–110
150. Vander Heiden MG, Cantley LC, Thompson CB (2009) Understanding the Warburg effect: the metabolic requirements of cell proliferation. Science 324:1029–1033
151. Vaupel P (2004) Tumor microenvironmental physiology and its implications for radiation oncology. Semin Radiat Oncol 14:198–206
152. Walenta S, Salameh A, Lyng H, Evensen JF, Mitze M, Rofstad EK, Mueller-Klieser W (1997) Correlation of high lactate levels in head and neck tumors with incidence of metastasis. Am J Pathol 150:409–415
153. Walenta S, Wetterling M, Lehrke M, Schwickert G, Sundfor K, Rofstad EK, Mueller-Klieser W (2000) High lactate levels predict likelihood of metastases, tumor recurrence, and restricted patient survival in human cervical cancers. Cancer Res 60:916–921
154. Wang R, Green DR (2012) Metabolic checkpoints in activated T cells. Nat Immunol 13:907–915
155. Wang JQ, Kon J, Mogi C, Tobo M, Damirin A, Sato K, Komachi M, Malchinkhuu E, Murata N, Kimura T et al (2004) TDAG8 is a proton-sensing and psychosine-sensitive G-protein-coupled receptor. J Biol Chem 279:45626–45633

156. Wang Z, Rong YP, Malone MH, Davis MC, Zhong F, Distelhorst CW (2006) Thioredoxin-interacting protein (txnip) is a glucocorticoid-regulated primary response gene involved in mediating glucocorticoid-induced apoptosis. Oncogene 25:1903–1913
157. Wang R, Dillon CP, Shi LZ, Milasta S, Carter R, Finkelstein D, McCormick LL, Fitzgerald P, Chi H, Munger J et al (2011) The transcription factor Myc controls metabolic reprogramming upon T lymphocyte activation. Immunity 35:871–882
158. Webb BA, Chimenti M, Jacobson MP, Barber DL (2011) Dysregulated pH: a perfect storm for cancer progression. Nat Rev Cancer 11:671–677
159. Wellen KE, Thompson CB (2010) Cellular metabolic stress: considering how cells respond to nutrient excess. Molecular cell 40:323–332
160. Wernicke CM, Richter GH, Beinvogl BC, Plehm S, Schlitter AM, Bandapalli OR, Prazeres da Costa O, Hattenhorst UE, Volkmer I, Staege MS et al (2012) MondoA is highly overexpressed in acute lymphoblastic leukemia cells and modulates their metabolism, differentiation and survival. Leuk Res 36:1185–1192
161. Williams AC, Collard TJ, Paraskeva C (1999) An acidic environment leads to p53 dependent induction of apoptosis in human adenoma and carcinoma cell lines: implications for clonal selection during colorectal carcinogenesis. Oncogene 18:3199–3204
162. Wise DR, DeBerardinis RJ, Mancuso A, Sayed N, Zhang XY, Pfeiffer HK, Nissim I, Daikhin E, Yudkoff M, McMahon SB et al (2008) Myc regulates a transcriptional program that stimulates mitochondrial glutaminolysis and leads to glutamine addiction. Proc Natl Acad Sci U S A 105:18782–18787
163. Wu N, Zheng B, Shaywitz A, Dragon Y, Tower C, Bellinger G, Chen C-H, Wen J, Asara JM, McGraw TE et al (2013) AMPK-dependent degradation of TXNIP upon energy stress leads to enhanced glucose uptake via GLUT1. Molecular cell 49:1–9
164. Yu FX, Goh SR, Dai RP, Luo Y (2009) Adenosine-containing molecules amplify glucose signaling and enhance txnip expression. Mol Endocrinol 23:932–942
165. Yuneva M, Zamboni N, Oefner P, Sachidanandam R, Lazebnik Y (2007) Deficiency in glutamine but not glucose induces MYC-dependent apoptosis in human cells. J Cell Biol 178:93–105
166. Zhou J, Chng WJ (2012) Roles of thioredoxin binding protein (TXNIP) in oxidative stress, apoptosis and cancer. Mitochondrion 13(3):163-169
167. Zhuo de X, Niu XH, Chen YC, Xin DQ, Guo YL, Mao ZB (2010) Vitamin D3 up-regulated protein 1(VDUP1) is regulated by FOXO3A and miR-17–5p at the transcriptional and post-transcriptional levels, respectively, in senescent fibroblasts. J Biol Chem 285:31491–31501
168. Zhou J, Yu Q, Chng WJ (2011) TXNIP (VDUP-1, TBP-2): a major redox regulator commonly suppressed in cancer by epigenetic mechanisms. Int J Biochem Cell Biol 43:1668–1673
169. Zoncu R, Bar-Peled L, Efeyan A, Wang S, Sancak Y, Sabatini DM (2011) mTORC1 senses lysosomal amino acids through an inside-out mechanism that requires the vacuolar H(+)-ATPase. Science 334:678–683

Chapter 6
Regulation of Renal Glutamine Metabolism During Metabolic Acidosis

Norman P. Curthoys

Overview of Glutamine Metabolism

Many rapidly growing cells such as intestinal epithelial cells and cells of the immune system consume glutamine as their primary nutrient [1]. These cells express high levels of a mitochondrial glutaminase (GA), which hydrolyzes the amide bond in glutamine. The resulting glutamate is transaminated to form either alanine or aspartate and α-ketoglutarate that is oxidized in the tricarboxylic acid cycle (TCA) cycle. As a result, these cells utilize very little glucose. By contrast, most cancer cells take up large amounts of glucose, which is converted primarily to lactate, even in the presence of sufficient oxygen [2]. To maintain mitochondrial function, transformed cells also exhibit an increased conversion of glutamine to lactate [3]. This pathway, termed glutaminolysis, generates adenosine triphosphate (ATP), reduced nicotinamide adenine dinucleotide phosphate (NADPH), and the precursors necessary to support the synthesis of the nucleotides and lipids, which are required for cell division. As a result, many transformed cells also require glutamine as an essential nutrient. To support this addiction to glutamine, transformed cells frequently exhibit an increased expression of the mitochondrial GA [4, 5].

Early studies used Ehrlich ascites tumor cells to demonstrate the role of GA in maintaining a transformed phenotype [6]. The tumor cells, which were stably transfected with a plasmid that encoded a GA antisense RNA, lost their transformed phenotype and failed to produce tumors when injected into mice. Interest in GA as a potential cancer chemotherapeutic target was kindled further by two recent studies. The initial study demonstrated that increased expression of the c-Myc oncogene in human P-493 B lymphoma cells resulted in increased expression of GA [7]. Further experiments demonstrated that c-Myc expression suppressed the synthesis of two microRNAs (miRNAs), miR-23a/b, that inhibit GA expression. The authors also demonstrated that small interfering RNA (siRNA) knockdown of GA significantly

N. P. Curthoys (✉)
Department of Biochemistry and Molecular Biology, Colorado State University,
Campus Delivery 1870, Fort Collins, CO 80523-1870, USA
e-mail: Norman.Curthoys@Colostate.edu

J-T. A. Chi (ed.), *Molecular Genetics of Dysregulated pH Homeostasis*,
DOI 10.1007/978-1-4939-1683-2_6

decreased the rates of proliferation of P493 B cells and human PC3 prostate cancer cells, two transformed cell lines that exhibit oncogenic levels of c-Myc. In the second study [8], compound 968 was identified as a potent inhibitor of the cellular transformation that was produced by expression of an oncogenic Dbl, a mutated form of a Rho-family guanine nucleotide exchange factor. Subsequent pull-down and mass spectrometric analysis identified the mitochondrial GA as the target of 968. Additional studies established that siRNA knockdown of GA mimicked the effects of 968. Both treatments inhibited the ability of three constitutively activated Rac or Rho GTPases to stimulate growth in low serum or in soft agar. GA knockdown also inhibited proliferation of transformed NIH3T3 fibroblasts and of breast cancer cells. More recent studies suggest that 968 is not a direct inhibitor of GA, but that it may block a covalent modification or the phosphate-dependent activation of GA [9].

Glutamine metabolism is also essential for the normal function of the nervous system. In the brain, GA participates in the intercellular cycle that generates and removes the excitatory neurotransmitter, glutamate [10]. This process normally maintains low micromolar concentrations of extracellular glutamate even in the presence of millimolar concentrations of extracellular glutamine. A slight increase in extracellular glutamate causes excitotoxicity and contributes to neuronal cell death resulting from stroke, trauma, and chronic neurodegenerative diseases [11–14]. Initial studies demonstrated that the release of GA from damaged neurons in cell culture results in the formation of extracellular glutamate and initiates a progression of excitotoxicity [15]. Additional studies demonstrated that following brain injury, high levels of released GA activity may contribute to the expanding zone of neuronal damage that evolves for 24–48 h after a stroke [16]. Thus, the gradual release of GA from disrupted neurons is a potential contributor to the in vivo development of glutamate excitotoxicity. Based upon these results, we designed a high throughput assay that was used by Robert Newcomb at Neurex to identify bis-2 [5-phenylacetamido-1,3,4-thiadiazol--yl] ethyl sulfide (BPTES) as a potent and specific inhibitor of GA. In subsequent studies, Jialin Zheng demonstrated that release of GA from HIV-infected macrophages [17] and microglia [18] contributes to HIV-associated dementia. The observed effects were prevented by BPTES inhibition or siRNA knockdown of GA [18–20]. Based upon these observations, it was proposed that a nonmembrane permeable inhibitor of the released GA would be an effective treatment to reduce the morbidity associated with a stroke or HIV-associated dementia. More recent studies have also demonstrated that BPTES selectively inhibits the growth of glioblastoma cells that express the R134 mutation in isocitrate dehydrogenase [21]. This mutation is frequently found in gliomas and acute myelogenous leukemia [22, 23]. The effect of BPTES was reproduced by selective siRNA knockdown of GA and was reversed by providing an exogenous source of α-ketoglutarate. Therefore, BPTES is a valuable lead compound for development of a therapeutic inhibitor of the mitochondrial GA.

The kidney is also an important site of glutamine metabolism, but only during metabolic acidosis [24]. During normal acid–base balance, the kidneys extract and catabolize very little of the plasma glutamine. However, following the onset of metabolic acidosis, which is characterized by a decrease in plasma pH and HCO_3^- ion concentration, the kidneys sustain a rapid and pronounced increase in glutamine catabolism. This adaptation occurs primarily in the proximal convoluted segment of the renal nephron and results in increased synthesis of NH_4^+ and HCO_3^- ions and glucose. The NH_4^+ ions are primarily excreted in the urine where they function as expendable cations, which facilitate the excretion of the excess acid. By contrast, the newly synthesized HCO_3^- ions are added to the blood to partially compensate the decrease in blood pH. The remaining carbons from glutamine are largely converted to glucose. This adaptation provides an important paradigm for understanding how GA expression and glutamine metabolism are regulated in response to onset of metabolic acidosis.

Structure of the Mitochondrial Glutaminase

Two separate genes, *GLS1* and *GLS2,* encode the mitochondrial GA. The human *GLS1* gene spans 82 kb on chromosome 2, contains 19 exons, and encodes two isoforms, kidney-type glutaminase (KGA) and glutaminase C (GAC), that are produced by alternative splicing of the initial transcript [25, 26]. The KGA isoform is highly expressed in the kidney, brain, intestine, and cells of the immune system, whereas GAC is primarily expressed in the heart, pancreas, and lung and is induced in many transformed cells. The KGA and GAC isoforms share identical N-terminal and core amino acid sequences that are transcribed from exons 1–14, but have unique C-termini (Fig. 6.1). The C-terminus of GAC is derived from exon 15, whereas the C-terminal domain of KGA is derived from exons 16–19 [27]. The C-termini of the KGA and GAC isoforms are nonhomologous but are highly conserved among various mammalian species. The N-terminal region of both rat isoforms contains a 72-amino acid mitochondrial targeting signal that is removed by the matrix processing protease [28]. The N-terminal segment of either mature *GLS1* protein contains a region of low complexity that is unstructured and highly sensitive to proteolysis [29]. The central core region, which contains the catalytic domain, is highly conserved from bacteria to humans. The Structural Genomics Consortium expressed this segment of the human *GLS1* gene product ($hGA_{221-533}$) in *Escherichia coli,* purified and crystallized it in the presence of glutamate, and solved the 3-dimensional structure by X-ray crystallography (http://www.rcsb.org/pdb/explore/explore.do?structureId=3CZD). This region forms a compact globular structure that is composed of two domains. One domain is entirely α-helical and the other contains both α-helices and β-sheets. The two domains form a pocket, which contains the active site serine residue. A co-crystallized glutamate molecule is tightly bound within this grove and is appropriately positioned adjacent to the active site serine residue.

By contrast, much less is known about the LGA isoform that is encoded by the *GLS2* gene [30]. This protein is highly expressed in adult liver and to a lesser extent in brain. It is also induced in multiple forms of cancer. LGA contains a catalytic

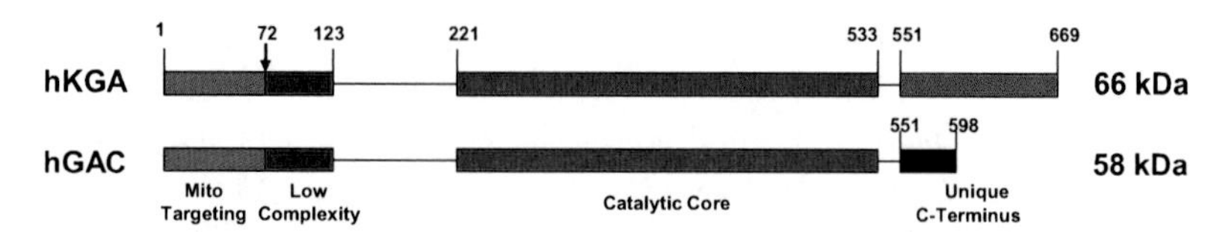

Fig. 6.1 Domain structure of human KGA and GAC isoforms. The two isoforms share a common sequence (amino acids 1–550) but contain unique C-terminal segments (KGA, amino acids 551–669 and GAC, amino acids 551–598)

core region that is highly homologous to the core regions of the *GLS1* proteins. However, it contains unique N- and C-terminal segments. The C-terminal region has been shown to bind a specific GA-interacting protein [31]. However, the function of this interaction has not been clearly defined.

A unique catalytic property of the KGA and GAC isoforms is the potent activation by phosphate and other polyvalent anions [32]. The K_M for glutamine decreases in the presence of increasing phosphate concentration and phosphate activation correlates with the association of inactive dimers to form active tetramers and larger oligomers [33, 34]. Kinetic and biophysical analyses of the mechanism of inactivation established that BPTES blocks the allosteric activation caused by phosphate binding and promotes the formation of an inactive complex [35]. Gel-filtration chromatography and sedimentation-velocity analysis established that BPTES prevents the formation of large phosphate-induced oligomers and instead promotes the formation of a single oligomeric species with distinct physical properties. Sedimentation-equilibrium studies determined that the oligomer produced by BPTES is a stable tetramer.

When expressed in *E. coli,* the full-length recombinant rat KGA forms inclusion bodies [36]. The N-terminal segment, containing the mitochondrial targeting signal and the region of low complexity, is encoded by exon 1. A construct that lacks this sequence ($rKGA_{\Delta 1}$) is readily expressed in *E. coli* and retains full activity [37]. This form of recombinant rKGA was used to perform the initial characterization of the effects of BPTES inhibition [35]. We obtained the $hGA_{221-533}$expression plasmid from the Structural Genomics Consortium and determined that this form of hGA is less active than the native enzyme. In addition, its activity is poorly inhibited by BPTES. To address these concerns, we cloned the $hKGA_{\Delta 1}$, $hGAC_{\Delta 1}$, and $hGA_{124-551}$ segments into pET-15b expression vectors, which add an N-terminal His_6-tag. The latter construct lacks the unique C-terminal sequences from either isoform. All three purified proteins retain full activity, indicating that neither C-terminal domain is essential for catalysis. Kinetic experiments established that the purified $hGA_{124-551}$protein exhibits a phosphate activation profile and a K_M for glutamine (2.5 mM) that are similar to the native enzyme [38]. When the glutamine saturation profiles were repeated using 30 mM phosphate in the presence of increasing concentrations of BPTES, both the K_M and the V_{MAX} decreased. Double reciprocal plots of the data indicated that BPTES functions as an uncompetitive inhibitor with a K_I of 0.2 μM. Therefore, BPTES is a potent inhibitor of hGA that binds to a site other than the active site and prevents a conformational change required for GA activity.

More recently, the structure of hGAC bound to BPTES was determined [39]. The full-length hGAC sequence was expressed in *Sf9* cells. During expression, the N-terminal mitochondrial targeting sequence was removed. When crystallized with BPTES, the hGAC forms a highly symmetrical tetramer containing two molecules of BPTES that are positioned at the dimer to dimer interfaces. The N-terminal regions of low complexity (residues 71–135) and the unique C-terminal segments (residues 547–598) were not evident in the X-ray crystallographic structure, suggesting that they are highly flexible or disordered. Residues 137–224 form small helical domains that are positioned on the sides of the tetramer opposite from the dimer to dimer interfaces. The catalytic core of the hGAC (residues 224–546) forms the highly conserved structure that was reported previously and is characteristic of all crystallized forms of GA. Each BPTES interacts in a highly symmetrical fashion with residues in the loop sequences (residues 320–327) and the α-helices (residues 386–399) that form the interface between two monomers. The loop sequence is normally unstructured, but it adopts a specific conformation upon binding of BPTES. The reported structure provides a detailed molecular model of how BPTES promotes the formation of an inactive tetrameric form of GA. The structure also suggests that the loop segment may play an important role in mediating the conformational changes that are essential for GA activity.

Renal Metabolism of Glutamine

During normal acid–base balance, the kidneys extract and metabolize very little of the plasma glutamine (Fig. 6.2). For example, the measured rat renal arterial–venous difference is less than 3 % of the arterial concentration of glutamine [40], and only 7 % of the plasma glutamine is extracted by the human kidneys even after an overnight fast [41]. Therefore, renal uptake is significantly less than the 20 % of plasma glutamine that is filtered by the glomeruli. Most of the filtered glutamine is reabsorbed within the proximal convoluted tubule and transported across the basolateral membrane [42]. Utilization of the small fraction of extracted plasma glutamine requires its transport into the mitochondrial matrix where glutamine is deamidated by KGA and then oxidatively deaminated by glutamate dehydrogenase (GDH). A mitochondrial glutamine transporter was partially purified from rat kidney and shown by reconstitution in lipid vesicles to be specific for glutamine and asparagine [43]. Neither the amino acid sequence nor the gene that encodes this transporter has been identified. However, kinetic measurements in isolated rat renal mitochondria indicate that the rate of glutamine uptake is not rate limiting for glutamine catabolism [44, 45]. In addition, basal GA activity is much greater than that required to accomplish the normal catabolism of glutamine. Therefore, either the activity of the mitochondrial glutamine transporter or GA must be largely inhibited or inactivated in vivo during normal acid–base balance to accomplish the effective reabsorption of glutamine. Finally, during normal acid–base balance, the urine is only slightly acidified. Thus, only two-thirds of the ammonium ions produced from

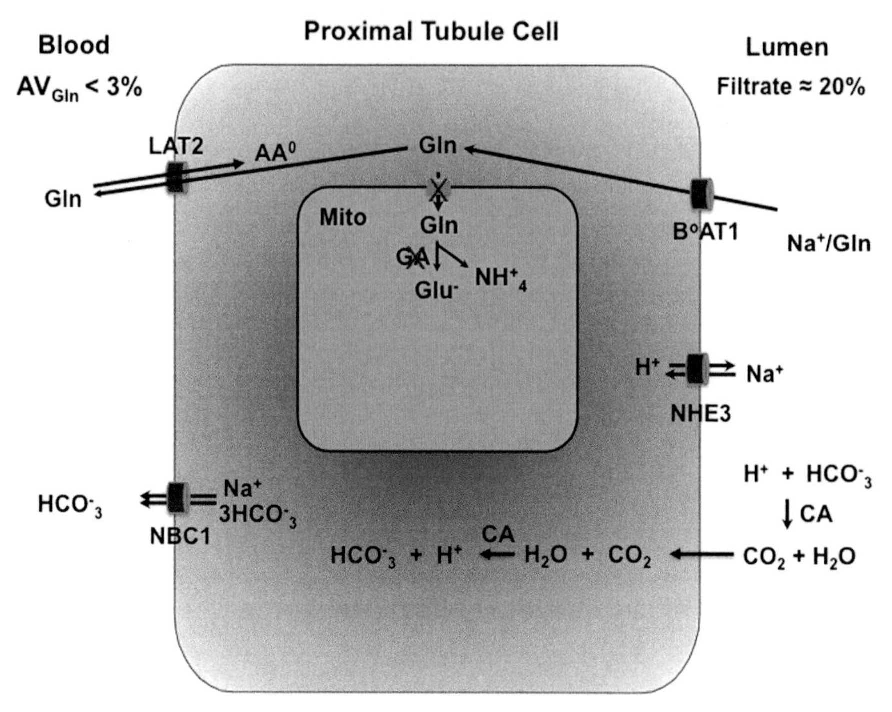

Fig. 6.2 Renal catabolism of glutamine during normal acid–base balance. The glutamine filtered by the glomeruli is nearly quantitatively extracted from the lumen of the proximal convoluted tubule and largely returned to the blood. The transepithelial transport utilizes *B°AT1*, a Na^+-dependent neutral amino acid cotransporter in the apical membrane, and *LAT2*, a neutral amino acid antiporter in the basolateral membrane. To accomplish this movement, either the mitochondrial glutamine transporter or the mitochondrial glutaminase (*GA*) must be inhibited (*red X*). The apical Na^+/H^+ exchanger functions to slightly acidify the lumen to facilitate the recovery of HCO_3^- ions

glutamine are trapped in the tubular lumen and excreted [46]. The remaining ammonium ions are added to the renal venous blood and utilized in the liver to generate urea, a process that consumes HCO_3^- [47].

Metabolic acidosis is a common clinical condition that is caused by genetic or acquired defects in metabolism, in renal handling of bicarbonate, and in the excretion of titratable acid [47]. In children, a sustained acidosis may contribute to the mental retardation that is frequently associated with many forms of genetic acidurias. Patients with cachexia, trauma, uremia, end-stage renal disease, and HIV infection frequently develop acidosis as a secondary complication that adversely affects outcome. In adults, chronic acidosis also contributes to osteomalacia, nephrocalcinosis, and urolithiasis.

Acute onset of a metabolic acidosis produces rapid changes in the interorgan metabolism of glutamine [48] that support a rapid and pronounced increase in renal catabolism of glutamine. Within 1–3 h, the arterial plasma glutamine concentration is increased twofold [49] due primarily to an increased release of glutamine from muscle tissue [50]. Significant renal extraction of glutamine becomes evident as the arterial plasma concentration is increased. Net extraction by the kidneys reach 35 % of the plasma glutamine, a level that significantly exceeds the 20 % that is filtered by the glomeruli. Thus, the direction of basolateral glutamine transport must be reversed in order for the proximal convoluted tubule cells to extract glutamine from both the glomerular filtrate and the venous blood. In addition, the transport of glutamine into the mitochondria and/or GA activity must be acutely activated [51]. Additional responses include a prompt acidification of the urine that results from translocation [52] and acute activation of NHE3, the apical Na^+/H^+ exchanger [53]. NHE3 can also transport NH_4^+ ions in place of H^+. Thus, the increased NHE3 activity also facilitates the rapid removal of cellular ammonium ions [54] and ensures that the bulk of the ammonium ions generated from the amide and amine nitrogens of glutamine are excreted in the urine. Finally, the cellular concentrations of glutamate and α-ketoglutarate are significantly decreased within the rat renal cortex [55]. The latter compounds are products and inhibitors of the GA and GDH reactions, respectively. The decrease in concentrations of the two regulatory metabolites may result from a pH-induced activation of α-ketoglutarate dehydrogenase [54]. Therefore, the acute increase in renal ammoniagenesis results from an increased availability of glutamine and the rapid activation of key transport processes and GA and GDH activities. All of these adaptations precede the increased expression of the enzymes of renal ammoniagenesis. Thus, the cells of the renal proximal convoluted tubule are likely to sense acute changes in extracellular or intracellular pH and/or HCO_3^- concentration and activate a signaling pathway that enhances flux through the mitochondrial KGA, GDH, and TCA cycle enzymes.

Potential pH-responsive Signaling Pathways

The kidney uses multiple mechanisms to sense changes in acid–base balance [56]. For example, GPR4, a G protein-coupled receptor that binds H^+ ions and activates cyclic adenosine monophosphate (cAMP) formation, is highly expressed throughout the renal collecting duct [57]. By sensing changes in interstitial pH, the GPR4 may stimulate cAMP production to activate V-ATPase translocation and H^+ ion extrusion by the A-type intercalated cells of the collecting duct [58]. A soluble adenyl cyclase (sAC), which is activated by HCO_3^- ions, is also highly expressed in the thick ascending limb, distal tubules, and collecting ducts of the kidney [59]. It may also function to activate V-ATPase translocation and H^+ ion excretion. However, neither of these proteins is significantly expressed in the proximal convoluted tubule where the increased expression of GA and GDH occur. In addition, treatment of LLC-PK_1-F^+ cells with forskolin or membrane-permeable analogs of cAMP had no effect on expression of the two key enzymes of renal ammoniagenesis [60].

The potential involvement of the MAPK signaling pathways (ERK1/2, SAPK/JNK, p38) in the pH-responsive induction of GA and phosphoenolpyruvate carboxykinase (PEPCK) mRNAs was initially examined by determining the effects of specific MAPK activators and inhibitors in LLC-PK_1-F^+ cells [61]. Anisomycin, a potent activator of the p38 and ERK1/2 pathways, increased GA and PEPCK mRNAs to levels comparable to those observed following treatment with pH 6.9 medium. Transfer of cultures to acidic medium resulted in phosphorylation, and thus activation, of both kinases. SB203580, a specific p38 MAPK inhibitor, led to a dose-dependent inhibition of both the pH- and anisomycin-mediated induction of PEPCK mRNA and blocked phosphorylation of activating transcription factor 2 (ATF-2), a downstream substrate of the p38 kinase. By contrast, the MEK1/2 inhibitor, PD098059, and the SAPK/JNK inhibitor, curcumin, did not affect basal or acid-induced levels of PEPCK mRNA, indicating that neither ERK1/2 nor SAPK/JNK play a significant role in regulating *PCK1* gene expression. Western blot analysis revealed that only p38α is strongly expressed in LLC-PK_1-F^+ cells. Gel-shift analysis using a labeled oligonucleotide containing the CRE-1 element of the PEPCK promoter produced multiple bands, one of which was supershifted with antibodies specific for ATF-2. Therefore, the p38α/ATF-2 signaling pathway is likely to mediate the pH-responsive induction of PEPCK mRNA. However, the addition of SB203580 did not block the pH-responsive increase in GA mRNA [61].

To further characterize the potential role of the p38 MAPK signaling pathway, clonal lines of LLC-PK_1-F^+ cells were developed that use a tetracycline-responsive promoter to express constitutively active (ca) or dominant negative (dn) forms of MKK3 and MKK6, kinases that act upstream of p38 MAPK [62]. Expression of caMKK6, but not caMKK3, caused an increase in phosphorylation of p38 MAPK and an increase in the level of PEPCK mRNA that closely mimicked the effect of treatment with acidic medium. The caMKK6 also enhanced expression of a *PCK1*-luciferase reporter construct. Co-expression of either or both dnMKKs blocked the increases in phosphorylation of p38 MAPK and transcription of PEPCK mRNA. Therefore, the pH-responsive increase in PEPCK mRNA in the kidney is mediated by the p38 MAPK signaling pathway and involves activation of MKK3 and/or MKK6 [62]. By contrast, co-expression of caMKK6 had no effect on the activity of a GA-luciferase construct that contained 401-bp of the GA promoter and the full-length 3′-UTR of the 3.4-kb KGA mRNA (unpublished data of C. Yang and N. P. Curthoys). The latter data indicate that separate signaling pathways are activated in response to acidosis to induce transcription of PEPCK mRNA and to stabilize selective mRNAs.

Response to Chronic Metabolic Acidosis

During chronic metabolic acidosis, the kidney continues to extract more than one third of the total plasma glutamine [40] in a single pass through this organ even though plasma glutamine concentrations are now decreased compared to normal and the initial decreases in glutamate and α-ketoglutarate are partially reversed. The increased renal catabolism of glutamine is now sustained by increased expression of the genes that encode various ion transporters and key enzymes of glutamine

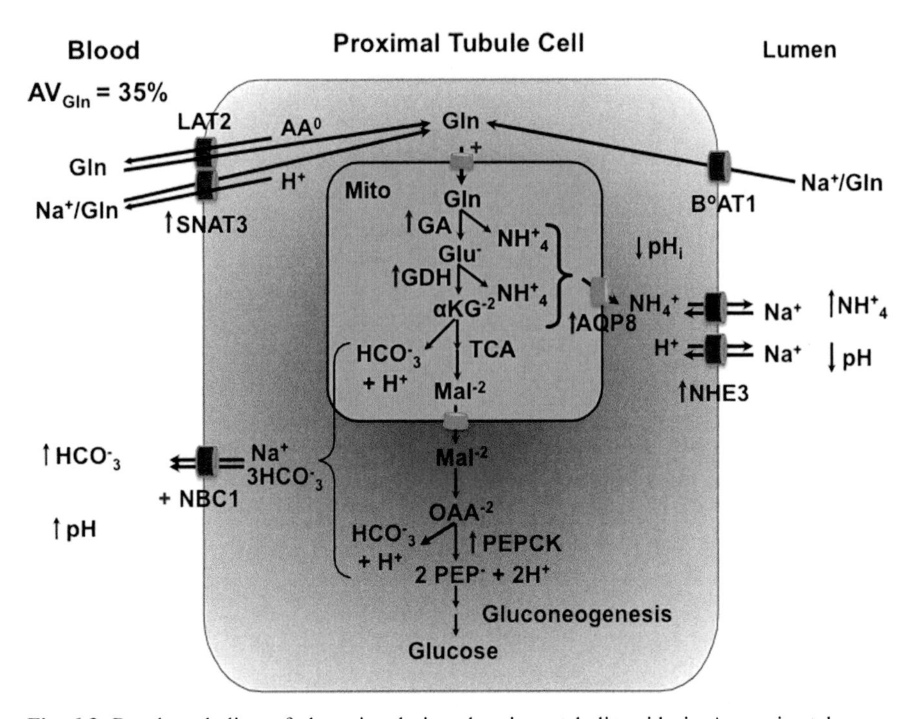

Fig. 6.3 Renal catabolism of glutamine during chronic metabolic acidosis. Approximately one-third of the plasma glutamine is extracted and catabolized within the early portion of the proximal tubule. *B°AT1* continues to mediate the extraction of glutamine from the lumen. Uptake of glutamine through the basolateral membrane occurs by reversal of the neutral amino acid exchanger, *LAT2*, and through increased expression of *SNAT3*. Increased renal catabolism of glutamine is facilitated by increased expression (*red arrows*) of the genes that encode glutaminase (GA), glutamate dehydrogenase (*GDH*), phospho*enol*pyruvate carboxykinase (*PEPCK*), the mitochondrial aquaporin-8 (*AQP8*), the apical Na^+/H^+ exchanger (NHE3), and the basolateral glutamine transporter (*SNAT3*). In addition, the activities of the mitochondrial glutamine transporter and the basolateral $Na^+/3HCO_3^-$ are increased (+). Increased expression of NHE3 contributes to the transport of ammonium ions and the acidification of the luminal fluid. The combined increases in renal ammonium ion excretion and gluconeogenesis result in a net synthesis of HCO_3^- ions that are transported across the basolateral membrane by the $Na^+/3HCO_3^-$ cotransporter (*NBC1*). *αKG* α-ketoglutarate, *Mal* malate, *OAA* oxaloacetate, *PEP* phosphoenolpyruvate

metabolism and gluconeogenesis (Fig. 6.3). Following the onset of acidosis, a rapid induction of PEPCK gene expression occurs only within the S1 and S2 segments of the proximal tubule [63]. The more gradual increases in the levels of mitochondrial GA [64, 65] and GDH [66] also occur solely within the proximal convoluted tubule. The adaptations in GA and PEPCK levels result from increased rates of synthesis of the proteins [67, 68] that correlate with comparable increases in the levels of their respective mRNAs [69, 70]. However, the increase in GA protein results from the selective stabilization of the GA mRNA [71–73], whereas the initial increase in PEPCK activity results from enhanced transcription of the *PCK1* gene [74]. However, the sustained increase in PEPCK expression is mediated, in part, by stabilization of the PEPCK mRNA [75, 76]. The activities of the mitochondrial glutamine transporter [51], SNAT3, a basolateral glutamine transporter [77], NHE3 [78], and NBC1, the basolateral Na^+-$3HCO_3^-$ cotransporter [78], are also increased in the proximal tubule during chronic acidosis.

SNAT3 (SN1, SLC38A3) [79] is a high-affinity glutamine transporter [80] that catalyzes the Na^+-dependent uptake of glutamine coupled to the efflux of an H^+ ion [81, 82]. Under physiological conditions, this reaction is reversible. In the brain, SNAT3 is expressed primarily in glial cells [83], while in the liver, it is found solely in the perivenous hepatocytes [84]. Under normal acid–base conditions, in rat kidney SNAT3 is localized solely to the basolateral membrane of the proximal straight tubules [77]. All of these cells express high levels of glutamine synthetase [1] and release glutamine. Thus, the SN1 transporter may normally function to catalyze glutamine efflux coupled to H^+ ion uptake. However, during acidosis, increased expression of SNAT3 occurs primarily in the basolateral membranes of the S1 and S2 segments of proximal tubule [79]. Given the sustained increases in H^+ ion concentration within these cells during acidosis [85, 86], the gradual increase in expression of the SN1 transporter may contribute to the sustained increase in basolateral uptake of glutamine.

The increased expression of NHE3 contributes to the acidification of the fluid in the tubular lumen and the active transport of ammonium ions [54]. Thus, the increased renal ammoniagenesis continues to provide an expendable cation that facilitates the excretion of titratable acids while conserving Na^+ and K^+ ions. The increased Na^+/H^+ exchanger activity also ensures the complete tubular reabsorption of HCO_3^- ions. In rats and humans, the α-ketoglutarate generated from glutamine is primarily converted to glucose [87]. This process requires the cataplerotic activity of PEPCK to convert intermediates of the TCA cycle to phospho*enol*pyruvate. This pathway generates two H^+ and two HCO_3^- ions per mole of α-ketoglutarate. The two H^+ ions are consumed during the conversion of phosphoenolpyruvate to glucose. The activation of NBC1, the basolateral $Na^+/3HCO_3^-$ cotransporter, facilitates the translocation of reabsorbed and of de novo-synthesized HCO_3^- ions into the renal venous blood. Thus, the combined adaptations create a net renal release of HCO_3^- ions that contribute to the ability of the kidney to partially restore acid–base balance.

Stabilization of GA mRNA

The time required for an mRNA to change from one steady-state level to another is proportional to its half-life [88]. Thus, rapid induction of an mRNA is feasible only if the mRNA has a rapid turnover. Eukaryotic mRNAs contain a 5′-^{7Me}GpppG cap and a 3′-poly(A) tail that bind initiation factors and the cytoplasmic poly(A) binding protein (PABP), respectively. Protein–protein interactions between eIF4E and PABP form a circular structure that stabilizes the mRNA and enhances translation. The rapid degradation of mammalian mRNAs can be initiated by the interaction of RNA binding proteins with specific sequences, usually AU-rich elements

(AREs), within the 3′-UTR [89, 90]. Alternatively, miRNAs, in complex with an Argonaute (AGO) protein, recognize specific mRNAs by base pairing to partially complementary binding sites [91]. The AGO protein interacts with GW182, which in turn interacts with the PABP. The primary mechanism of mammalian mRNA decay involves recruitment of either or both of the resulting ribonucleoprotein (RNP) complexes to cytoplasmic loci termed processing bodies. The processing bodies contain the CCR4/CAF1/Not deadenylase that removes the poly(A) tail. The deadenylated mRNA subsequently undergoes decapping by the Dcp1/Dcp2 complex and degradation by Xrn1, a 5′→ 3′ exonuclease [92].

We initially demonstrated that LLC-PK$_1$-F$^+$ cells, a gluconeogenic subline of porcine proximal tubule-like cells [93], model the pH-responsive increase in GA mRNA [94]. When transferred to an acidic medium (pH 6.9, 9 mM HCO_3^-), the cells exhibit a threefold increase in GA mRNA. Initial studies established that an increase in GA mRNA was observed with more physiological changes in pH and that the observed increase was proportional to the decrease in medium pH [94]. Therefore, pH 6.9 media was adopted to maximize the signal observed in future experiments. The presence of a sequence element that regulates the turnover of the GA mRNA was initially demonstrated by stable expression of various β-globin (βG) reporter mRNAs [71]. Expression of the parent βG construct in LLC-PK$_1$-F$^+$ cells produced a high level of a very stable mRNA ($t_{1/2}$ > 30 h). The turnover of the βG mRNA was not affected by transfer of the cells to acidic medium. A second construct, pβG-GA, included a 956-bp 3′-UTR segment that was derived from the rat KGA mRNA. The βG-GA mRNA exhibits a decreased expression that results from its more rapid turnover ($t_{1/2}$ = 4.6 h). Transfer of the cells expressing this construct to acidic medium (pH 6.9, 10 mM HCO_3^-) resulted in a pronounced stabilization and a gradual induction of the βG-GA mRNA. These studies indicated that the 3′-nontranslated segment contains a pH-response element (pHRE).

We subsequently mapped the pHRE through functional analysis of various deletion constructs and by characterizing specific RNA binding interactions [73]. The 956-nt segment was initially divided into three segments of nearly equal length, termed R-1, R-2, and R-3. They were cloned individually and in different combinations into the βG reporter construct and tested for their ability to enhance turnover and pH-responsiveness of the chimeric mRNA. The combined effects were retained in the 340-nt R-2 segment. RNA electrophoretic mobility shift assays indicated that cytosolic extracts of rat renal cortex contain a protein that binds to the R-2 RNA. The observed binding was mapped by deletion analysis to a 29-nt fragment, termed R-2I, which contains a direct repeat of two 8-nt AU sequences. The binding interaction was reduced significantly by mutating either 8-nt element and was completely lost by mutating both elements. The binding was effectively competed by an excess of the same RNA, but not by adjacent or unrelated RNAs. UV-cross-linking experiments labeled a 35-kDa protein that binds to the R-2I RNA. Thus, a protein binds specifically to two 8-nt AU-elements within the 3′-UTR of the GA mRNA.

We further characterized the specificity and functional significance of the observed binding interactions by measuring the effect of acidic medium on the half-lives of various chimeric βG-GA construct [72]. Insertion of short segments of GA mRNA containing the direct repeat or a single 8-nt AU-sequence was sufficient to impart a fivefold pH-responsive stabilization to the chimeric mRNA. Furthermore, site-directed mutation of the direct repeats of the 8-nt AU-sequence in a βG-GA mRNA completely abolished the pH-responsive stabilization. Thus, either the direct repeat or a single 8-nt AU-sequence is both necessary and sufficient to create a functional pHRE.

To determine if deadenylation precedes the degradation of the GA mRNA, a tetracycline-responsive promoter system was developed in LLC-PK_1-F^+ cells to perform a pulse-chase analysis of the turnover of βG-GA mRNA [95]. This system uses cells that stably express high levels of the tTA transcription factor, a chimeric protein that contains a tetracycline-response element-binding domain and a VP16 activation domain. These cells were then stably transfected with various chimeric βG reporter constructs that contain a minimal promoter with multiple tetracycline-response elements. In the absence of doxycycline, the tTA protein binds to the promoter and greatly enhances transcription of the chimeric βG mRNA. When added at concentrations >50 ng/ml, doxycycline binds to the tTA protein and inhibits transcription. Thus, this approach accomplishes the rapid induction and shut off of the synthesis of a single mRNA and avoids indirect effects that are caused by the use of a general transcription inhibitor [96]. With this approach, the measured half-life of the βG-GA mRNA is 2.9 h in LLC-PK_1-F^+ cells maintained in normal medium and is increased fivefold when the cells are transferred to acidic medium. RNase H cleavage and Northern analysis of the 3′-ends established that rapid deadenylation occurred concomitant with the rapid decay of the βG-GA mRNA in cells grown in normal medium. Stabilization of the βG-GA mRNA in cells treated with acidic medium is associated with a pronounced decrease in the rate and the extent of deadenylation. The data indicate that deadenylation is the initial and potentially rate-limiting step in the turnover of the GA mRNA. Mutation of the pHRE within the βG-GA mRNA blocked the pH-responsive stabilization but not the rapid degradation, whereas insertion of only a 29-bp segment containing the pHRE was sufficient to produce both rapid degradation and pH-responsive stabilization. Therefore, the 3′-UTR of the GA mRNA must contain additional instability elements. However, the identified pHRE contributes to the rapid turnover of the GA mRNA and is both necessary and sufficient to mediate its pH-responsive stabilization [95].

During chronic metabolic acidosis, the adaptive increase in rat renal ammoniagenesis is also sustained, in part, by increased expression of GDH [66, 97]. The 3′-UTR of the GDH mRNA [98] contains four 8-base sequences that are 88 % identical to one of the two pHREs present in the GA mRNA. Insertion of the 3′-UTR of the GDH complementary DNA (cDNA) into the βG expression vector produced a chimeric mRNA that was stabilized threefold when LLC-PK_1-F^+ cells were transferred to acidic medium. A similar pH-responsive stabilization was also observed using a βG construct that contained only GDH4, one of the four AREs. Therefore, during acidosis, pH-responsive stabilization of the GDH mRNA may be accomplished by the same mechanism that affects an increase in the GA mRNA [98].

The pHRE from the GA mRNA was used as an affinity ligand to identify ζ-crystallin, an NADPH quinone reductase, as the primary protein in rat kidney cortical extracts that interacts with this RNA segment [99]. Further experiments with highly purified recombinant protein confirmed that ζ-crystallin binds to the pHRE with high affinity and specificity. Therefore, an adenovirus construct was used to overexpress mouse ζ-crystallin in the LLC-PK_1-F^+ cells that stably express the βG-GA reporter construct from the tet-responsive promoter. However, a 50-fold overexpression of ζ-crystallin had no effect on the basal half-life or the pH-responsive stabilization of the βG-GA mRNA [100]. In addition, a porcine-specific ζ-crystallin shRNA was stably expressed in LLC-PK_1-F^+ cells. This reduced expression of ζ-crystallin to a nondetectable level, <2% of the level in nontransformed cells. The reduced expression of ζ-crystallin had no effect on the endogenous levels of GAC, KGA, or PEPCK. Thus, in spite of the fact that ζ-crystallin is the primary protein in extracts of rat kidney cortex [99] and of porcine LLC-PK_1-F^+ proximal tubule cells [95] that binds to the pHRE, it now appears unlikely that ζ-crystallin contributes to the rapid degradation or the selective stabilization of the GA mRNA.

Therefore, the potential of three other well-characterized ARE binding proteins, p40 AU-binding factor 1 (p40AUF1) [101], TTP [102], and human antigen R (HuR) [103] to interact with the pHRE was characterized. The highly purified recombinant proteins form specific complexes with the pHRE. AUF1 and HuR function as dimers and are known to form multiple complexes in an RNA gel-shift assay. For each of the proteins, a concentration between 0.1 and 0.4 μM is required to shift 50% of the labeled RNA. Therefore, the three recombinant proteins may bind to the pHRE with similar affinities. Western blot analyses demonstrated that rat kidney cortex and cultured kidney cells express HuR and all four isoforms of AUF1 [95], but failed to detect TTP expression. However, Brfl and Brf2, two homologs of TTP [104], are expressed at high levels in rat kidney cortex. Thus, one or more of these proteins may contribute to the regulation of GA mRNA turnover.

Stabilization of PEPCK mRNA

Previous studies established that PEPCK mRNA is degraded rapidly in liver and hepatoma cells [105] and in rat kidney cortex [106]. The tetracycline-responsive promoter system was used to accurately quantify the half-lives of various chimeric β-globin-PEPCK (βG-PCK) mRNAs in LLC-PK_1-F^+ cells [107]. Characterization of cells that stably express the βG-PCK-1 mRNA, which contains the entire 3′-UTR of the PEPCK mRNA established that this mRNA is degraded with a very rapid half-life ($t_{1/2}$=2.1 h). RNase H treatment of βG-PCK-1 mRNA established that rapid deadenylation and mRNA degradation occur concomitantly. The half-lives of various deletion constructs were quantified in order to map the elements that mediate the rapid decay of the PEPCK mRNA. The βG-PCK-2 mRNA, containing the

5′-end of the 3′-UTR, was degraded with a half-life of 5.4 h. By contrast, βG-PCK-3 mRNA that contains the 3′-half of 3′-UTR was degraded more rapidly ($t_{½}=3.6$ h). The βG-PCK-6/7 mRNA, which contains only 73 nucleotides from the 3′-end of the 3′-UTR of PEPCK mRNA has the same half-life, indicating that all of the instability elements in the 3′-end of the 3′-UTR are contained in this segment.

AUF1 is a well-characterized RNA-binding protein that usually enhances mRNA turnover [108]. It was initially identified as an enhancer of ARE-mediated decay of c-myc mRNA in extracts of K562 cells [109]. AUF1 is usually expressed as four isoforms, p37, p40, p42, and p45, which are produced by differential splicing of exons 2 and 7 from the initial transcript of the *AUF1* gene [110]. All four isoforms contain two RNA-binding motifs and exhibit similar sequence specific binding. The two larger isoforms incorporate a C-terminal 49 amino acid insertion from exon 7 that blocks nuclear export. As a result, the p42 and p45 isoforms are largely retained within the nucleus. By contrast, the p37 and p40, which differ by an N-terminal 19 amino acid insertion from exon 2, shuttle between the nucleus and the cytosol. Thus, the two smaller isoforms are more likely to affect mRNA stability. However, LLC-PK_1-F^+ cells express very low levels of p37AUF1. Thus, RNA gel-shift analyses were performed using purified recombinant p40AUF1. The results established that p40AUF1 binds with high affinity and specificity to the PCK-2, PCK-6, and PCK-7 segments of the 3′-UTR of PEPCK mRNA [107]. Mutational analysis indicated that p40AUF1 binds to a highly conserved UUAUUUAU sequence within PCK-6 and to a stem-loop structure and adjacent CU-region in PCK-7. Thus, AUF1 binds to multiple destabilizing elements within the 3′-UTR. Furthermore, the observed number of interactions closely correlates with the rate of turnover of the various βG-PCK mRNAs. Therefore, the binding of AUF1 may contribute to the rapid turnover of the PEPCK mRNA.

The transfer of LLC-PK_1-F^+ cells to an acidic medium (pH 6.9, 9 mM HCO_3^-) produced an increased expression of PEPCK mRNA that occurred following a pronounced delay and reached a 2.5-fold maximum after 18 h. However, this increase in expression occurred with no evident change in the half-life of the PEPCK mRNA [111]. The LLC-PK_1-F^+ cells used in previous studies were a mixed population of cells. Thus, clonal lines of LLC-PK_1-F^+ cells were selected to identify a cell line that exhibits a greater fold increase in cytosolic PEPCK mRNA and protein [76]. When treated with acidic medium, the clonal LLC-PK_1-F^+-9C cells exhibit a more rapid and more pronounced increase in PEPCK mRNA and protein that reached a four- to fivefold increase after 15 and 20 h, respectively. Measurement of the half-lives established that the endogenous PEPCK mRNA turns over rapidly ($t_{½}=3.2$ h) in cells treated with normal medium (pH 7.4, 26 mM HCO_3^-), but is stabilized twofold when the cells are transferred to acidic medium. The pH-responsive stabilization was reproduced by the Tet-responsive expression of βG-PCK-1 mRNA. Therefore, the clonal line of LLC-PK_1-F^+ cells effectively models both the transcriptional activation and the pH-responsive stabilization of renal PEPCK mRNA. The latter response was lost by mutation of the ARE within the PCK-6 segment. This segment contains a 17-nucleotide AU-sequence which has a high degree of identity to the 16-nucleotide AU-sequence that mediates the pH-responsive stabilization of the KGA mRNA [72]. In addition, 11 of the 17-nucleotides (UUAAAUUAUUU) are fully conserved within the 3′-end of the 3′-UTR of all of the mammalian *PCK1* genes that have been sequenced. Disruption of this AU-sequence by introduction of seven G and C nucleotides resulted in a twofold stabilization of the βG-PCK-1 mRNA and prevented a further stabilization of the chimeric mRNA when the cells were transferred to an acidic medium. The PCK-6 segment also binds AUF1 and is the primary element that mediates the rapid turnover of PEPCK mRNA. Therefore, this highly conserved sequence is the primary element that mediates the rapid turnover and the pH-responsive stabilization of the PEPCK mRNA.

HuR is another well-characterized ARE-binding protein [103] that is ubiquitously expressed [112]. It contains three conserved RNA-binding domains that belong to the RNA recognition motif (RRM) superfamily [113]. The first and second RRMs bind with high affinity to AREs, while the third binds to the poly(A) tail [114, 115]. HuR also contains a 33 amino acid hinge region between the second and third RNA-binding domains which functions as a nucleocytoplasmic shuttling sequence [116]. Through identification of its target transcripts, HuR has been implicated in the control of cell division, carcinogenesis, immune responsiveness, and the response to various cellular stresses [117]. Electrophoretic mobility shift assays established that purified recombinant HuR also binds with high affinity and specificity to two sites within the 3′-UTR of the PEPCK mRNA [76]. These sites overlap with the AUF1-binding sites in the PCK-6 and PCK-7 segments (Fig. 6.4). siRNA knockdown of HuR in LLC-PK$_1$-F$^+$-9C cells caused a pronounced decrease in basal expression and reduced the pH-responsive increases in PEPCK mRNA and protein. Most importantly, the siRNA knockdown of HuR also blocked the pH-responsive increase in the half-life of the endogenous PEPCK mRNA. However, treatment with acidic medium had no effect on the level or subcellular distribution of HuR or the various isoforms of AUF1 [76]. Therefore, the pH-responsive stabilization of PEPCK mRNA may require covalent modifications of HuR and/or AUF1, which affect their binding to the elements that mediate the rapid turnover of PEPCK mRNA.

siRNA knockdown of HuR in LLC-PK$_1$-F$^+$-9C cells also prevented the pH-responsive increase in PEPCK mRNA half-life, suggesting that HuR is necessary for this response [75]. A recruitment assay, using a reporter mRNA in which the pHRE of the PEPCK 3′-UTR were replaced with six MS2 stem-loop sequences, was developed to test the hypothesis. The individual recruitment of a chimeric protein containing the MS2 coat protein and either HuR or p40AUF1 failed to produce a pH-responsive stabilization. However, the concurrent expression of both chimeric proteins was sufficient to produce a pH-responsive increase in the half-life of the reporter mRNA. siRNA knockdown of AUF1 produced slight increases in basal levels of PEPCK mRNA and protein, but partially inhibited the pH-responsive increases. Complete inhibition of the latter response was achieved by knockdown of both RNA-binding proteins. The results suggest that binding of HuR and AUF1 have opposite effects on basal expression, but they may interact to mediate the pH-responsive increase in PEPCK mRNA. Two-dimensional gel electrophoresis indicated that treatment with an acidic medium caused a decrease in phosphorylation of HuR, but may increase phosphorylation of the multiple AUF1 isoforms. Thus, the pH-responsive stabilization of PEPCK mRNA requires the concurrent binding of HuR and AUF1 and may be mediated by changes in their extent of covalent

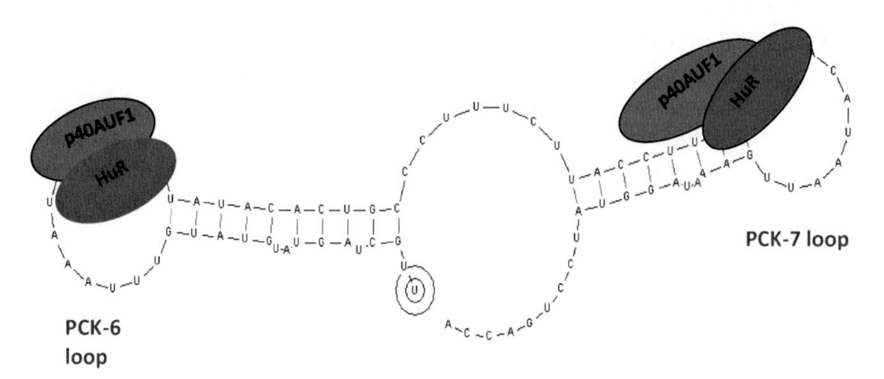

Fig. 6.4 HuR and p40AUF1 bind to the same stem-loop structures at the 3′-end of the 3′-UTR of rat PEPCK mRNA. The PCK-6 and PCK-7 segments of the 3′-UTR of PEPCK mRNA form stem-loop structures in which the loops contain highly conserved AU-rich sequences that bind HuR and p40AUF1 with high affinity and specificity

modification. Given the rapid turnover and demonstrated stabilization, PEPCK mRNA expression in the clonal LLC-PK$_1$-F$^+$-9C cells provides an excellent model system and effective paradigm to further characterize the molecular mechanism that mediates a major component of the renal response to acidosis.

Conclusions

Characterization of the sustained increases in the expression of the mitochondrial GA and GDH and the cytosolic PEPCK during metabolic acidosis has defined a novel mechanism of pH-responsive mRNA stabilization. The characterization of the initial increase in PEPCK gene expression has also served as a paradigm to characterize the mechanism of pH-responsive changes in transcription. In addition, some insight has been gained in identifying the signal transduction pathways that sense changes in pH and HCO_3^- ion concentration and mediate this adaptive response in the kidney. However, given the complexity of this process, the expression of a large number of additional genes are likely to also respond to changes in acid–base balance. Further understanding of the complexity and the regulation of this important physiological process may be derived by the application of genomic, proteomic, and metabolomic techniques to identify the full set of pathways that are differentially regulated in the proximal convoluted tubule. The resulting data will identify the full spectra of genes that are activated or repressed by metabolic acidosis and differentiate those that are regulated through transcriptional or posttranscriptional mechanisms. A thorough systems analysis may also identify sets of genes, and their associated regulatory elements, that are temporally affected by mechanisms that have not, as yet, been characterized. Such techniques may also be used to identify and more fully characterize the signal transduction pathways that mediate and coordinate the overall regulation of gene expression during metabolic acidosis.

References

1. Taylor L, Curthoys NP (2004) Glutamine metabolism: role in acid-base balance. Biochem Mol Biol Ed 32:291–304
2. Warburg O, Wind F, Negelein E (1927) The metabolism of tumors in the body. J Gen Physiol 8(6):519–530
3. DeBerardinis RJ et al (2008) The biology of cancer: metabolic reprogramming fuels cell growth and proliferation. Cell Metab 7(1):11–20
4. DeBerardinis RJ et al (2007) Beyond aerobic glycolysis: transformed cells can engage in glutamine metabolism that exceeds the requirement for protein and nucleotide synthesis. Proc Natl Acad Sci U S A 104(49):19345–19350
5. Erickson JW, Cerione RA (2010) Glutaminase: a hot spot for regulation of cancer cell metabolism? Oncotarget 1(8):734–740
6. Segura JA et al (2001) Ehrlich ascites tumor cells expressing anti-sense glutaminase mRNA lose their capacity to evade the mouse immune system. Int J Cancer 91(3):379–384
7. Gao P et al (2009) c-Myc suppression of miR-23a/b enhances mitochondrial glutaminase expression and glutamine metabolism. Nature 458(7239):762–765
8. Wang JB et al (2010) Targeting mitochondrial glutaminase activity inhibits oncogenic transformation. Cancer Cell 18(3):207–219
9. Katt WP et al (2012) Dibenzophenanthridines as inhibitors of glutaminase C and cancer cell proliferation. Mol Cancer Ther 11(6):1269–1278
10. Masson J et al (2006) Mice lacking brain/kidney phosphate-activated glutaminase have impaired glutamatergic synaptic transmission, altered breathing, disorganized goal-directed behavior and die shortly after birth. J Neurosci 26(17):4660–4671
11. Meldrum BS (1994) The role of glutamante in epilepsy and other CNS disorders. Neurology 44(Suppl 8):S14–S23
12. Choi DW, Rothman SM (1990) The role of glutamate neurotoxicity in hypoxic-ischemic neuronal death. Annu Rev Neurosci 13(1):171–182
13. Muir KW, Lees KR (1995) Clinical experience with excitatory amino acid antagonist drugs. Stroke 26:503–513
14. Rothman SM, Olney JW (1986) Glutamate and the pathopysiology of hypoxic-ischemic brain damage. Ann Neurol 2:105–111
15. Newcomb R et al (1997) Increased production of extracellular glutamate by the mitochondrial glutaminase following neuronal death. J Biol Chem 272(17):11276–11282
16. Newcomb R et al (1998) Characterization of mitochondrial glutaminase and amino acids at prolonged times after experimental focal cerebral ischemia. Brain Res 813(1):103–111
17. Zhao J et al (2004) Mitochondrial glutaminase enhances extracellular glutamate production in HIV-1-infected macrophages: linkage to HIV-1 associated dementia. J Neurochem 88(1):169–180
18. Huang Z et al (2011) Glutaminase dysregulation in HIV-1-infected human microglia mediates neurotoxicity: relevant to HIV-1 associated neurocognitive disorders. J Neurosci 31:15195–15204
19. Erdmann N et al (2009) In vitro glutaminase regulation and mechanisms of glutamate generation in HIV-1-infected macrophage. J Neurochem 109(2):551–561
20. Erdmann N et al (2007) Glutamate production by HIV-1 infected human macrophage is blocked by the inhibition of glutaminase. J Neurochem 102(2):539–549
21. Seltzer MJ et al (2010) Inhibition of glutaminase preferentially slows growth of glioma cells with mutant IDH1. Cancer Res 70(22):8981–8987
22. Dang L et al (2009) Cancer-associated IDH1 mutations produce 2-hydroxyglutarate. Nature 462(7274):739–744
23. Ward PS et al (2010) The common feature of leukemia-associated IDH1 and IDH2 mutations is a neomorphic enzyme activity converting alpha-ketoglutarate to 2-hydroxyglutarate. Cancer Cell 17(3):225–234

24. Curthoys NP (2012) Renal ammonium ion production and excretion. In: Alpern RJ, Moe OW, Caplan M (eds) The kidney: physiology and pathophysiology, 5th edn. Elsevier, San Diego, pp 1993–2018
25. Elgadi KM, et al (1999) Cloning and analysis of unique human glutaminase isoforms generated by tissue-specific alternative splicing. Physiol Genomics 1(2):51–62
26. Porter D et al (1995) Differential expression of multiple glutaminase mRNAs in LLC-PK1-F + cells. Am J Physiol 269(3 Pt 2):F363–F373.
27. Porter LD et al (2002) Complexity and species variation of the kidney-type glutaminase gene. Physiol Genomics 9(3):157–166
28. Srinivasan M, Kalousek F, Curthoys NP (1995) In vitro characterization of the mitochondrial processing and the potential function of the 68-kDa subunit of renal glutaminase. J Biol Chem 270(3):1185–1190
29. Haser WG, Shapiro RA, Curthoys NP (1985) Comparison of the phosphate-dependent glutaminase obtained from rat brain and kidney. Biochem J 229(2):399–408
30. Perez-Gomez C et al (2003) Genomic organization and transcriptional analysis of the human L-glutaminase gene. Biochem J 370(Pt 3):771–784
31. Olalla L et al (2001) The C-terminus of human glutaminase L mediates association with PDZ domain-containing proteins. FEBS Lett 488(3):116–122
32. Curthoys NP, Kuhlenschmidt T, Godfrey SS (1976) Regulation of renal ammoniagenesis. Purification of phosphate-dependent glutaminase from rat kidney. Arch Biochem Biophys 174:82–89
33. Godfrey SS, Kuhlenschmidt T, Curthoys NP (1977) Correlation between activation and dimer formation of rat renal phosphate-dependent glutaminase. J Biol Chem 252:1927–1931
34. Morehouse RF, Curthoys NP (1981) Properties of rat renal phosphate-dependent glutaminase coupled to sepharose. Evidence that dimerization is essential for activation. Biochem J 193:709–716
35. Robinson MM et al (2007) Novel mechanism of inhibition of rat kidney-type glutaminase by bis-2-(5-phenylacetamido-1,2,4-thiadiazol-2-yl)ethyl sulfide (BPTES). Biochem J 406(3):407–414
36. Shapiro RA et al (1991) Isolation, characterization, and in vitro expression of a cDNA that encodes the kidney isoenzyme of the mitochondrial glutaminase. J Biol Chem 266(28):18792–18796
37. Kenny J et al (2003) Bacterial expression, purification, and characterization of rat kidney-type mitochondrial glutaminase. Protein Expr Purif 31(1):140–148
38. Hartwick E, Curthoys N (2012) BPTES inhibition of hGA124–551, a truncated form of human kidney-type glutaminase. J Enzyme Inhib Med Chem 27:861–867 (15 Oct 2011 Epub)
39. DeLaBarre B et al (2011) Full-length human glutaminase in complex with an allosteric inhibitor. Biochemistry 50(50):10764–10770
40. Squires EJ, Hall DE, Brosnan JT (1976) Arteriovenous differences for amino acids and lactate across kidneys of normal and acidotic rats. Biochem J 160(1):125–128
41. Meyer C et al (2002) Renal substrate exchange and gluconeogenesis in normal postabsorptive humans. Am J Physiol Endocrinol Metab 282(2):E428–E434
42. Silbernagl S (1980) Tubular reabsorption of L-glutamine studied by free-flow micropuncture and microperfusion of rat kidney. Int J Biochem 12(1–2):9–16
43. Indiveri C et al (1998) Identification and purification of the reconstitutively active glutamine carrier from rat kidney mitochondria. Biochem J 333(Pt 2):285–290
44. Kovacevic Z, Bajin K (1982) Kinetics of glutamine-efflux from liver mitochondria loaded with the 14C-labeled substrate. Biochim Biophys Acta 687(2):291–295
45. Goldstein L, Boylan JM (1978) Renal mitochondrial glutamine transport and metabolism: studies with a rapid-mixing, rapid-filtration technique. Am J Physiol 234(6):F514–F521
46. Sleeper RS et al (1978) Effects of acid challenge on in vivo and in vitro rat renal ammoniagenesis. Life Sci 22(18):1561–1571
47. Halperin ML (1993) Metabolic aspects of metabolic acidosis. Clin Invest Med 16(4):294–305

48. Tamarappoo BK, Joshi S, Welbourne TC (1990) Interorgan glutamine flow regulation in metabolic acidosis. Miner Electrolyte Metab 16(5):322–330
49. Hughey RP, Rankin BB, Curthoys NP (1980) Acute acidosis and renal arteriovenous differences of glutamine in normal and adrenalectomized rats. Am J Physiol 238(3):F199–F204
50. Schrock H, Cha CJ, Goldstein L (1980) Glutamine release from hindlimb and uptake by kidney in the acutely acidotic rat. Biochem J 188(2):557–560
51. Sastrasinh M, Sastrasinh S (1990) Effect of acute pH change on mitochondrial glutamine transport. Am J Physiol 259(6 Pt 2):F863–F866
52. Yang X et al (2000) Acid incubation causes exocytic insertion of NHE3 in OKP cells. Am J Physiol Cell Physiol 279(2):C410–C419
53. Horie S et al (1990) Preincubation in acid medium increases Na/H antiporter activity in cultured renal proximal tubule cells. Proc Natl Acad Sci U S A 87(12):4742–4745
54. Tannen RL, Ross BD (1979) Ammoniagenesis by the isolated perfused rat kidney: the critical role of urinary acidification. Clin Sci (Lond) 56(4):353–364
55. Lowry M, Ross BD (1980) Activation of oxoglutarate dehydrogenase in the kidney in response to acute acidosis. Biochem J 190(3):771–780
56. Brown D, Wagner CA (2012) Molecular mechanisms of acid-base sensing by the kidney. J Am Soc Nephrol 23(5):774–780
57. Sun X et al (2010) Deletion of the pH sensor GPR4 decreases renal acid excretion. J Am Soc Nephrol 21(10):1745–1755
58. Gong F et al (2010) Vacuolar H+-ATPase apical accumulation in kidney intercalated cells is regulated by PKA and AMP-activated protein kinase. Am J Physiol Renal Physiol 298(5):F1162–F1169
59. Pastor-Soler N et al (2003) Bicarbonate-regulated adenylyl cyclase (sAC) is a sensor that regulates pH-dependent V-ATPase recycling. J Biol Chem 278(49):49523–49529
60. Liu X, Curthoys NP (1996) cAMP activation of phosphoenolpyruvate carboxykinase transcription in renal LLC-PK1-F+ cells. Am J Physiol 271(2 Pt 2):F347–F355
61. Feifel E et al (2002) p38 MAPK mediates acid-induced transcription of PEPCK in LLC-PK(1)-FBPase(+) cells. Am J Physiol Renal Physiol 283(4):F678–F688
62. O'Hayre M et al (2006) Effects of constitutively active and dominant negative MAPK kinase (MKK) 3 and MKK6 on the pH-responsive increase in phosphoenolpyruvate carboxykinase mRNA. J Biol Chem 281(5):2982–2988
63. Guder WG, Ross BD (1984) Enzyme distribution along the nephron. Kidney Int 26(2):101–111
64. Curthoys NP, Lowry OH (1973) The distribution of glutaminase isoenzymes in the various structures of the nephron in normal, acidotic, and alkalotic rat kidney. J Biol Chem 248(1):162–168
65. Wright PA, Knepper MA (1990) Phosphate-dependent glutaminase activity in rat renal cortical and medullary tubule segments. Am J Physiol 259(6 Pt 2):F961–F970
66. Wright PA, Knepper MA (1990) Glutamate dehydrogenase activities in microdissected rat nephron segments: effects of acid-base loading. Am J Physiol 259(1 Pt 2):F53–F59
67. Iynedjian PB, Ballard FJ, Hanson RW (1975) The regulation of phosphoenolpyruvate carboxykinase (GTP) synthesis in rat kidney cortex. The role of acid-base balance and glucocorticoids. J Biol Chem 250(14):5596–5603
68. Tong J, Harrison G, Curthoys NP (1986) The effect of metabolic acidosis on the synthesis and turnover of rat renal phosphate-dependent glutaminase. Biochem J 233(1):139–144
69. Cimbala MA et al (1982) Rapid changes in the concentration of phosphoenolpyruvate carboxykinase mRNA in rat liver and kidney. Effects of insulin and cyclic AMP. J Biol Chem 257(13):7629–7636
70. Hwang JJ, Curthoys NP (1991) Effect of acute alterations in acid-base balance on rat renal glutaminase and phosphoenolpyruvate carboxykinase gene expression. J Biol Chem 266(15):9392–9396
71. Hansen WR et al (1996) The 3′-nontranslated region of rat renal glutaminase mRNA contains a pH-responsive stability element. Am J Physiol 271(1 Pt 2):F126–F131

72. Laterza OF, Curthoys NP (2000) Specificity and functional analysis of the pH-responsive element within renal glutaminase mRNA. Am J Physiol Renal Physiol 278(6):F970–F977
73. Laterza OF et al (1997) Identification of an mRNA-binding protein and the specific elements that may mediate the pH-responsive induction of renal glutaminase mRNA. J Biol Chem 272(36):22481–22488
74. Hanson R, Reshef L (1997) Regulation of phosphoenolpyruvate carboxykinase (GTP) gene expression. Annu Rev Biochem 66:581–611
75. Gummadi L, Taylor L, Curthoys NP (2012) Concurrent binding and modifications of AUF1 and HuR mediate the pH-responsive stabilization of phosphoenolpyruvate carboxykinase mRNA in kidney cells. Am J Physiol Renal Physiol 303(11):F1545–F1554
76. Mufti J et al (2011) Role of AUF1 and HuR in the pH-responsive stabilization of phosphoenolpyruvate carboxykinase mRNA in LLC-PK-F cells. Am J Physiol Renal Physiol 301(5):F1066–F1077
77. Karinch AM et al (2002) Regulation of expression of the SN1 transporter during renal adaptation to chronic metabolic acidosis in rats. Am J Physiol Renal Physiol 283(5):F1011–F1019
78. Preisig PA, Alpern RJ (1988) Chronic metabolic acidosis causes an adaptation in the apical membrane Na/H antiporter and basolateral membrane Na(HCO3)3 symporter in the rat proximal convoluted tubule. J Clin Invest 82(4):1445–1453
79. Solbu TT et al (2005) Induction and targeting of the glutamine transporter SN1 to the basolateral membranes of cortical kidney tubule cells during chronic metabolic acidosis suggest a role in pH regulation. J Am Soc Nephrol 16(4):869–877
80. Christensen HN (1990) Role of amino acid transport and countertransport in nutrition and metabolism. Physiol Rev 70(1):43–77
81. Chaudhry FA, Reimer RJ, Edwards RH (2002) The glutamine commute: take the N line and transfer to the A. J Cell Biol 157(3):349–355
82. Bode BP (2001) Recent molecular advances in mammalian glutamine transport. J Nutr 131(Suppl 9):2475S–2485S
83. Chaudhry FA et al (1999) Molecular analysis of system N suggests novel physiological roles in nitrogen metabolism and synaptic transmission. Cell 99(7):769–780
84. Gu S et al (2000) Identification and characterization of an amino acid transporter expressed differentially in liver. Proc Natl Acad Sci U S A 97(7):3230–3235
85. Ackerman JJ et al (1981) The role of intrarenal pH in regulation of ammoniagenesis: [31P] NMR studies of the isolated perfused rat kidney. J Physiol 319:65–79
86. Adam WR, Koretsky AP, Weiner MW (1986) 31P-NMR in vivo measurement of renal intracellular pH: effects of acidosis and K+ depletion in rats. Am J Physiol 251(5 Pt 2):F904–F910
87. Gerich JE et al (2001) Renal gluconeogenesis: its importance in human glucose homeostasis. Diabetes Care 24(2):382–391
88. Hargrove JL (1993) Microcomputer-assisted kinetic modeling of mammalian gene expression. FASEB J 7(12):1163–1170
89. Chen CY, Shyu AB (1995) AU-rich elements: characterization and importance in mRNA degradation. Trends Biochem Sci 20(11):465–470
90. Parker R, Song H (2004) The enzymes and control of eukaryotic mRNA turnover. Nat Struct Mol Biol 11(2):121–127
91. Huntzinger E, Izaurralde E (2011) Gene silencing by microRNAs: contributions of translational repression and mRNA decay. Nat Rev Genet 12(2):99–110
92. Anderson P, Kedersha N (2006) RNA granules. J Cell Biol 172(6):803–808
93. Gstraunthaler G, Handler JS (1987) Isolation, growth, and characterization of a gluconeogenic strain of renal cells. Am J Physiol 252(2 Pt 1):C232–C238
94. Kaiser S, Curthoys NP (1991) Effect of pH and bicarbonate on phosphoenolpyruvate carboxykinase and glutaminase mRNA levels in cultured renal epithelial cells. J Biol Chem 266(15):9397–9402
95. Schroeder JM et al (2006) Role of deadenylation and AUF1 binding in the pH-responsive stabilization of glutaminase mRNA. Am J Physiol Renal Physiol 290(3):F733–F740

96. Loflin PT et al (1999) Transcriptional pulsing approaches for analysis of mRNA turnover in mammalian cells. Methods 17(1):11–20
97. Curthoys NP, Gstraunthaler G (2001) Mechanism of increased renal gene expression during metabolic acidosis. Am J Physiol Renal Physiol 281(3):F381–F390
98. Schroeder JM, Liu W, Curthoys NP (2003) pH-responsive stabilization of glutamate dehydrogenase mRNA in LLC-PK1-F+ cells. Am J Physiol Renal Physiol 285(2):F258–F265
99. Tang A, Curthoys NP (2001) Identification of zeta-crystallin/NADPH:quinone reductase as a renal glutaminase mRNA pH response element-binding protein. J Biol Chem 276(24):21375–21380
100. Ibrahim H, Lee YJ, Curthoys NP (2008) Renal response to metabolic acidosis: role of mRNA stabilization. Kidney Int 73(1):11–18
101. Wilson GM, Brewer G (1999) The search for trans-acting factors controlling messenger RNA decay. Prog Nucleic Acid Res Mol Biol 62:257–291
102. Blackshear PJ (2002) Tristetraprolin and other CCCH tandem zinc-finger proteins in the regulation of mRNA turnover. Biochem Soc Trans 30(Pt 6):945–952
103. Brennan CM, Steitz JA (2001) HuR and mRNA stability. Cell Mol Life Sci 58(2):266–277
104. Phillips RS, Ramos SB, Blackshear PJ (2002) Members of the tristetraprolin family of tandem CCCH zinc finger proteins exhibit CRM1-dependent nucleocytoplasmic shuttling. J Biol Chem 277(13):11606–11613
105. Hod Y, Hanson RW (1988) Cyclic AMP stabilizes the mRNA for phosphoenolpyruvate carboxykinase (GTP) against degradation. J Biol Chem 263(16):7747–7752
106. Hanson RW, Patel YM (1994) Phosphoenolpyruvate carboxykinase (GTP): the gene and the enzyme. Adv Enzymol Relat Areas Mol Biol 69:203–281
107. Hajarnis S, Schroeder JM, Curthoys NP (2005) 3′-untranslated region of phosphoenolpyruvate carboxykinase mRNA contains multiple instability elements that bind AUF1. J Biol Chem 280(31):28272–28280
108. Zucconi BE, Wilson GM (2011) Modulation of neoplastic gene regulatory pathways by the RNA-binding factor AUF1. Front Biosci 17:2307–2325
109. Zhang W et al (1993) Purification, characterization, and cDNA cloning of an AU-rich element RNA-binding protein, AUF1. Mol Cell Biol 13(12):7652–7665
110. Wagner BJ et al (1998) Structure and genomic organization of the human AUF1 gene: alternative pre-mRNA splicing generates four protein isoforms. Genomics 48(2):195–202
111. Holcomb T et al (1996) Promoter elements that mediate the pH response of PCK mRNA in LLC-PK1-F+ cells. Am J Physiol 271(2 Pt 2):F340–F346
112. Ma WJ et al (1996) Cloning and characterization of HuR, a ubiquitously expressed Elav-like protein. J Biol Chem 271(14):8144–8151
113. Burd CG, Dreyfuss G (1994) Conserved structures and diversity of functions of RNA-binding proteins. Science 265(5172):615–621
114. Abe R et al (1996) Two different RNA binding activities for the AU-rich element and the poly(A) sequence of the mouse neuronal protein mHuC. Nucleic Acids Res 24(24):4895–4901
115. Ma WJ, Chung S, Furneaux H (1997) The elav-like proteins bind to AU-rich elements and to the poly(A) tail of mRNA. Nucleic Acids Res 25(18):3564–3569
116. Fan XC, Steitz JA (1998) HNS, a nuclear-cytoplasmic shuttling sequence in HuR. Proc Natl Acad Sci U S A 95(26):15293–15298
117. Gallouzi IE, Brennan CM, Steitz JA (2001) Protein ligands mediate the CRM1-dependent export of HuR in response to heat shock. RNA 7(9):1348–1361

Chapter 7
Extracellular Acidosis and Cancer

Maike D. Glitsch

Solid tumours acidify their interstitial fluid by a number of diverse processes that result in net proton export. These include activity of monocarbonate and bicarbonate transporters, V-type ATPases, carbonic anhydrases and voltage-gated proton channels. The acidic extracellular microenvironment influences the activity of cancer, vascular endothelial and immune cells such that tumour growth is promoted although the exact signalling pathways are still only poorly understood. This chapter addresses how changes in extracellular proton concentration can affect cells in cancerous tissue.

Extracellular Acidification of Tumour Interstitial Fluid

Acidification of the interstitial fluid is a hallmark of solid tumours. Cancerous cells use glycolysis for energy production, which results in accumulation of lactate and it was originally thought that the extrusion of this through monocarboxylate transporters was the main course for the low extracellular pH. Acidification of the extracellular fluid was hence regarded as a consequence of altered tumour metabolic activity (which is thought to be adopted by cancer cells to allow them to continue adenosine triphosphate (ATP) production even under anoxic conditions) and that cancer cells somehow managed to cope with. However, findings that glycolytically inactive cancer cells still acidify the extracellular fluid [11, 27, 44] suggest that other proton extrusion mechanisms contribute to the accumulation of protons outside of cancer cells. These include sodium–proton exchangers, bicarbonate transporters and plasmalemmar V-type proton ATPases as well as carbonic anhydrases that are catalytically active on the extracellular side and catalyse the formation of carbonic acid from CO_2 and H_2O; the bicarbonate is then taken back up into the cell, leaving the protons behind (reviewed by Swietach et al. [35]). Moreover, recent

M. D. Glitsch (✉)
Department of Physiology, Anatomy and Genetics, University of Oxford, Oxford, England
e-mail: maike.glitsch@dpag.ox.ac.uk

J-T. A. Chi (ed.), *Molecular Genetics of Dysregulated pH Homeostasis*,
DOI 10.1007/978-1-4939-1683-2_7

publications report the presence of voltage-gated proton channels (Hv1) in highly metastatic breast tumour tissue [40], the expression of which correlated with tumour size and acidification of the tumour tissue [41]. Hv1 channels are voltage-gated ion channels, the voltage dependence of which is shifted towards more hyperpolarised values with increasing intracellular acidification, resulting in their opening and allowing diffusion of protons out of the cell under resting conditions upon intracellular acidification [7]. These findings suggest that Hv1 channels can contribute to the acidification of the interstitial fluid observed in cancerous tissues. Hence, there are numerous proton extrusion mechanisms that cancer cells can utilise to acidify the extracellular medium. Cancer cells are in fact so good at exporting protons that the cytoplasmic pH (pH_i) of cancer cells is not only more alkaline than in normal cells but also more alkaline than the extracellular pH (pH_o). Crucially, this represents an inversion of proton distribution in cancerous cells compared with healthy, non-transformed cells, in which the pH_i to pH_o ratio is 7.2:7.4, whereas in cancer cells, it may be 7.4:6.0, although more commonly pH_o values of around 6.5–7.0 are reported [26, 42].

Intriguingly, it appears that acidification of the interstitial fluid is a requirement for tumour progression and spread. First, injection of human melanoma cells that had been cultured under acidic (pH_o 6.8) and control (pH_o 7.4) conditions into athymic nude mice showed that cells cultured under acidic conditions displayed an increased invasive, angiogenic and metastatic potential than cells cultured at pH_o 7.4 [31]. Second, acid conditioning of melanoma cells led to the generation of more invasive cancer cells with altered gene expression, suggesting that extracellular acidosis affects gene expression and that these changes in gene expression are required to promote metastatic potential of the cells [23]. Third, treatment of tumour-bearing mice with pH buffers (bicarbonate or the nonvolatile pH buffer IEPA) reduced the metastatic activity of tumours [10, 30].

Hence, it would appear that extracellular acidosis is a desired outcome, rather than merely tolerated by cancer cells. Addressing the question of why cancer cells can thrive in an acidic environment when normal, non-transformed cells cannot is therefore of utmost importance, since therapies interfering with tumour acidification and its consequences are likely to provide useful clinical strategies for combating this disease family. It is hence important to understand the impact that an increase in extracellular proton concentration has on cell function in both the short- and long-term. The short-term impact is relevant for understanding how extracellular acidity can affect protein function and gene expression immediately such that it protects cancer cells from the detrimental impact of high extracellular proton concentrations observed in most nontransformed cells, whereas the long-term impact allows us to identify which proteins/genes are required to keep cancer cells functioning optimally in this hostile environment. Immediate and long-term effects are likely to be mediated by distinct genes/proteins and hence represent distinct potential targets for treatment strategies.

Influence of Extracellular Acidosis on Plasma Membrane Protein Function

Generally speaking, extracellular protons can affect a cell by two independent mechanisms that both have the potential to profoundly influence the cell's ability to relate information from its membrane surface to distinct intracellular compartments as well as impact on its ability to communicate with other cells. First, protons influence protein structure. Amongst physiologically relevant ions, protons are the smallest and hence have the largest charge density. This, together with their small size, means that they can powerfully interfere with protein structure [22]. Their presence may disrupt salt bridges formed between oppositely charged amino acid residues of a given protein, thereby altering the three-dimensional structure and hence function of that protein. Importantly, external acidosis may inhibit or potentiate protein function. It is likely that transmembrane proteins of the plasma membrane are particularly affected since their cytosolic and extracellular components will be exposed to the different proton concentrations inside and out, and it is conceivable that an alkalinisation of pH_i concomitant with an acidification of pH_o can synergistically affect transmembrane protein function. Moreover, extracellular protons can act as ligands in their own right that bind to proton-sensing proteins and thereby activate them, thus communicating the presence of protons (i.e. a low pH_o) to the cytosol. Proton-gated transmembrane proteins include ion channels (acid-sensing ion channels, ASICs) and transient receptor potential channels of the vanilloid family (TRPV channels), particularly member 1 (TRPV1 [8, 12]) and G protein-coupled receptors (ovarian cancer gene 1, OGR1), G protein-coupled receptor 4 (GPR4) and T cell death associated gene 8 (TDAG8) [8, 12, 33]. Finally, extracellular acidosis may also impact on the activity of extracellular enzymes. Thrombin, a hydrolase involved in blood coagulation, is inhibited by external acidosis [3], whereas cathepsin B, which is normally a lysosomal protease but which is secreted or translocated to the cell surface in tumour cells, requires an acidic pH to function; intriguingly, these proteases have been implicated in a number of tumour promoting processes [21].

Second, protons are charge carriers and as such can influence the membrane potential. Virtually all cells have negative membrane potentials, meaning that the cytosol is negatively charged with respect to the extracellular fluid (which is taken as electrically neutral). This membrane potential is due to the unequal distribution of ions across the cell's membrane and the membrane's selective permeability to these ions. The distribution of protons in normal tissue counters this negative membrane potential because the proton concentration inside is higher than outside (pH_o 7.4 versus pH_i 7.2). An increase in extracellular proton concentration (as found in cancerous tissue) may hence lead to hyperpolarisation of the membrane potential with important consequences for cell function, as the membrane potential influences the activity of transmembrane proteins directly and indirectly. A direct influence of membrane potential on transmembrane protein function is observed in a number of distinct protein families: voltage-gated ion channels, voltage-sensing enzymes [37], voltage-sensitive G protein-coupled receptors (GPCRs) [2, 19, 20], voltage-sensing transporter proteins [25, 28] and voltage-gated transmembrane motor proteins [45]. These proteins have intrinsic voltage sensors (usually charged amino acids) that respond to a change in membrane potential by altering protein structure, thereby allowing them to change their activity in a membrane potential dependent manner. Voltage-gated ion channels are directly gated in their activity by membrane potential, whereas the activity of GPCRs and transporter proteins tends to be modulated (inhibited or potentiated) rather than directly gated by the membrane potential.

Membrane potential can also indirectly influence transmembrane protein activity, with ion channels and electrogenic transporter proteins as key targets. Ion channels and electrogenic transporters can influence the membrane potential because they permit the flow of ions across the otherwise impermeable membrane, thereby altering the distribution of charges across the membrane. Their function is, in turn, affected by the membrane potential because it, in part, determines the driving force (a combination of electrical and chemical gradient that determines the direction of net ionic flow for a given ion) for ionic movement across the plasma membrane: A negative membrane potential potentiates net anion efflux and net cation influx, whereas a more depolarised membrane potential has the opposite effect.

In this context, the impact of extracellular acidosis on intracellular Ca^{2+} signalling in cancer cells is of particular interest. Cytoplasmic Ca^{2+} concentrations are kept at very low levels (typically between 10 and 100 nM), and changes in Ca^{2+} homeostasis trigger changes in cellular processes including modulation of enzyme activity, ion transport protein function and gene transcription. Ultimately, changes in cytoplasmic Ca^{2+} concentration can affect cellular processes such as cell cycle progression, proliferation, migration and apoptosis. These processes can be altered in cancerous cells to suit transformed cell requirements (for recent reviews, see [6, 24]), and it is therefore important to understand if and how extracellular acidosis links to intracellular Ca^{2+} signalling pathways.

Cytoplasmic Ca^{2+} can be increased by two distinct pathways: Ca^{2+} influx through Ca^{2+} permeable plasma membrane proteins (ion channels and ion pumps/transporters) or Ca^{2+} release from intracellular Ca^{2+} stores (with Ca^{2+} again either diffusing through Ca^{2+} permeable channels or being transported by pumps/transporters from the store into the cytoplasm). Crucially, transport proteins/pumps that remove cytoplasmic Ca^{2+} under physiological conditions may contribute to increases in cytoplasmic Ca^{2+} concentrations when their function is reversed or inhibited. One such example is the depletion of the intracellular Ca^{2+} store that is the endoplasmic reticulum (ER): The ER contains high concentrations of Ca^{2+} (values vary between 12 μM and 2 mM for free Ca^{2+} concentration within the store; variation is due to different methods used to determine the intraluminal Ca^{2+} concentration and due to distinct preparations/cell types used in the experiments [4]. The ER membrane expresses both a Ca^{2+} leak conductance that permits Ca^{2+} to continually escape from the ER as well as a powerful Ca^{2+} pump (SERCA pump—sarcoplamic endoplamic reticulum calcium ATPase) that pumps Ca^{2+}, which has left the ER via the leak pathway, back into the ER. This may seem like a futile cycle but, like leak channels and Na^+/K^+ pumps are crucially involved in setting up and fine-tuning the membrane potential of any given cell, this Ca^{2+} leak and Ca^{2+} pump cycle allows fine tuning

Table 7.1 Effects of extracellular acidosis on plasma membrane proteins

	H^+ activated	H^+ potentiated	H^+ inhibited
G protein-coupled receptor	OGR1 TDAG8 GPR4	$P2Y_4$	CaSR[a] mGluR4[b]
Ion channel	ASIC1-4 TRPV1	TRPC4β and 5 TRPM7 $P2X_2$ homomer $P2X_{2+3}$ heteromer $P2X_3$ homomer TREK2 GIRK1/4 heteromer $K_v1.3$	TRPC6 TRPV6 $P2X_7$[c] TASK-1, -2, -3 TRESK TWIK-1 TALK-1, -2 $K_{ir}1.1$, 4.1, 5.1 $K_v1.4$, 1.5, 11.1 $Ca_v3.1$

References for this table are taken from Glitsch [8] and Holzer [12], unless otherwise indicated
[a]Quinn et al. [29]
[b]Levinthal et al. [16]
[c]Virginio et al. [38]

of cytoplasmic Ca^{2+} levels by altering the activity of the SERCA pumps and/or leak pathway. Inhibition of the SERCA pump using the pharmacological inhibitor thapsigargin is a commonly used tool to deplete the ER of its Ca^{2+} content, underlining the importance and potency of this pathway.

Changes in extracellular pH can affect both Ca^{2+} influx and Ca^{2+} release from intracellular Ca^{2+} stores by (1) acting directly on proton-sensitive, Ca^{2+} permeable ion channels and pumps in the plasma membrane that permit passage of Ca^{2+} across the plasma membrane, (2) affecting proton-sensitive receptor proteins that couple to intracellular Ca^{2+} stores through activation of the phospholipase C pathway or regulate the function of ion channels/pumps in the plasma membrane and thereby indirectly impact on cytoplasmic Ca^{2+} levels, or (3) by affecting K^+ channels that are involved in setting the membrane potential, thereby at least, in part, determining the driving force for Ca^{2+} ions to enter the cytoplasm. A number of proton-sensitive ion channels have been identified whose activity can be gated (i.e. activated), potentiated (i.e. enhanced but not triggered) or inhibited by increases in extracellular proton concentration. These include nonselective cation channels with varying degrees of Ca^{2+} permeability, Ca^{2+} selective ion channels and K^+ selective ion channels, which are important when considering changes in Ca^{2+} influx in response to extracellular acidification. Interestingly, all extracellular proton-gated channels identified to date are nonselective cation channels. These channels can affect cytoplasmic Ca^{2+} levels by depolarising the membrane potential, thereby promoting voltage-dependent Ca^{2+} influx but decreasing voltage-independent pathways (since a depolarisation decreases the driving force for Ca^{2+}), or by directly permitting Ca^{2+} to enter the cytoplasm. K^+ selective channels hyperpolarise the membrane potential, thus promoting voltage-independent Ca^{2+} influx by increasing the driving force for Ca^{2+} influx but inhibiting depolarisation-dependent pathways. A list of proton-sensitive plasma membrane GPCRs and ion channels is given in Table 7.1; this table is by no means exhaustive but it illustrates clearly that protons can impact differentially on distinct members of the same (super) family of membrane proteins. It also demonstrates that distinct channels that are permeable to the same kind of ion can be differently affected by protons. Potassium channels are an excellent paradigm to exemplify this: voltage-dependent $K_v1.3$ channels are potentiated by extracellular acidosis whereas voltage-dependent $K_v1.4$, 1.5 and 11.1 channels are not. Similarly, two pore domain potassium (K2P) channels can be both potentiated and inhibited by extracellular acidification.

Another crucial aspect to bear in mind when considering the impact of extracellular proton concentration on cell function is that different proteins exhibit different sensitivities to protons, which is shown in Fig. 7.1a. The threshold for activation of proton-potentiated/activated and inhibited membrane proteins can vary quite significantly: Some proteins are already partially active (inhibited) as physiological pH (e.g. OGR1 [18]) whereas others require a much more pronounced drop in extracellular pH (e.g. pH6 for TRPV1 activation [36]). Additionally, the impact of extracellular pH may be time dependent (Fig. 7.1b) since some proton-sensing membrane proteins may desensitise to the prolonged presence of extracellular protons (e.g. ASIC1a is nearly fully desensitised within 10 s [39]), whereas others do not (e.g. TRPV1 [36]; OGR1 [18]). Hence, there is a whole arsenal of distinct plasma membrane proteins that not only respond differently to extracellular acidosis in terms of level of activation, potentiation or inhibition at a given extracellular pH but also in terms of kinetics of the response. The nature and number of proton-sensitive plasma membrane proteins will therefore critically shape the kind of response that a cell displays in the presence of external acidosis. Cells can modify their response to extracellular acidosis by changing gene expression, which allows them to express proteins that are either more or less sensitive to extracellular acidosis and/or monitor proton concentrations in either the short-term or long-term.

There is quite a substantial body of literature linking ion channels and GPCRs to tumour formation and cancer progression, and many of these proteins are sensitive to changes in extracellular proton concentrations (Table 2 in [8]). Crucially, extracellular acidosis can not only influence cancer cells and render them more aggressive [23, 31], but also impact on noncancerous cells that are equally found in cancerous tissues, including vascular endothelial cells and immune cells, and promote cancer progression. Whilst it is well documented *that* interstitial acidosis promotes tumour progression, it is less well understood *how* an increase in extracellular proton concentration achieves this effect. In human medulloblastoma cells (derived from a pediatric cerebellar brain tumour), extracellular acidosis was shown to activate the ERK pathway via activation of OGR1 and subsequent increases in intracellular Ca^{2+} concentration [13], providing a mechanistic explanation on how an increase in intracellular proton concentration can influence gene transcription in cancer cells. Crucially, these receptors do not desensitise [18] and hence faithfully report extracellular proton concentrations at all times.

OGR1 and GPR4 have also been implicated in angiogenesis in tumour tissue. Hypervascularity is a common feature of solid tumours and angiogenesis is required to allow tumours to exceed sizes of around 2 mm (set by the diffusion limit for oxygen and nutrients). In OGR1 knock-out mice, injection of melanoma B16-F10

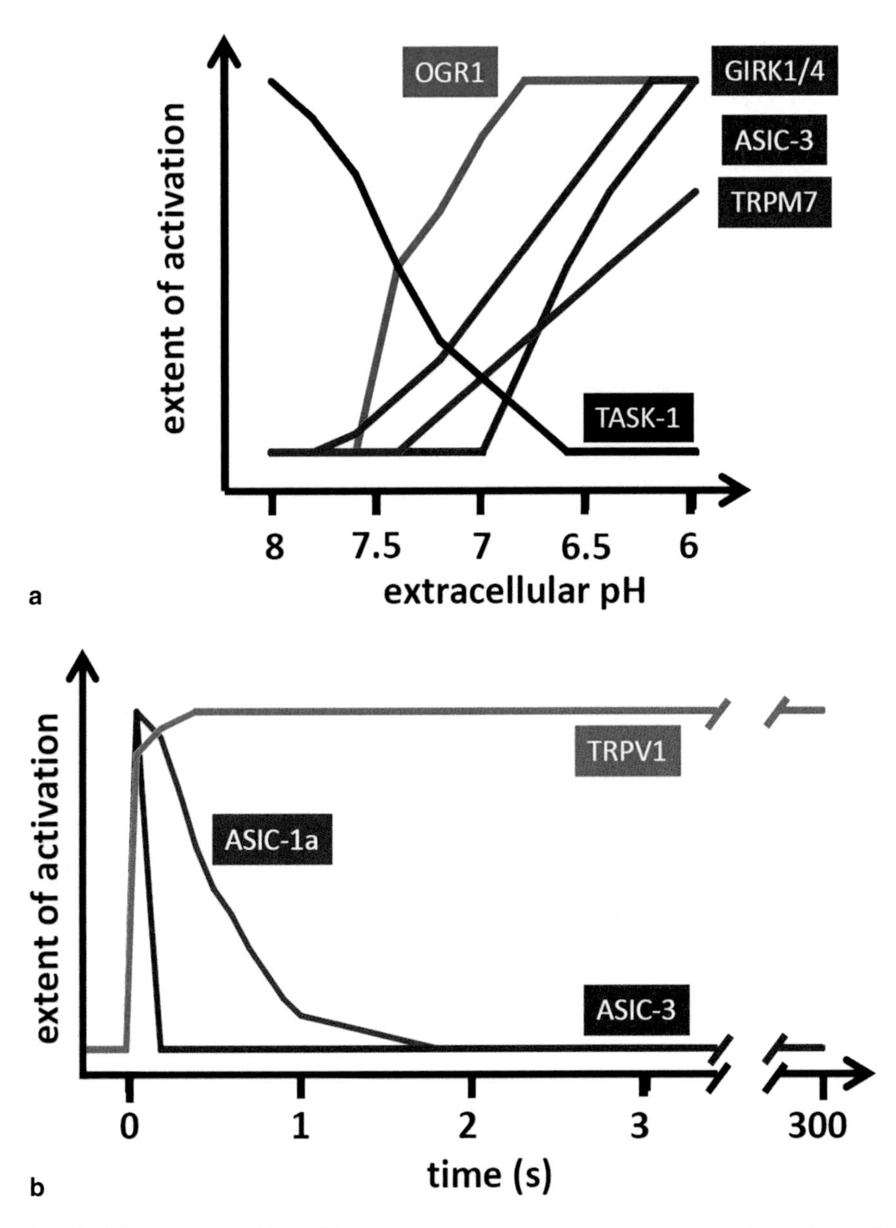

Fig. 7.1 Effect of extracellular acidosis on plasma membrane protein function. **a** Dependence of plasma membrane protein function on the extent of external acidosis. *TASK-1* channels are inhibited with increasing proton concentration, whilst *GIRK1/4* and *TRPM7* channels are potentiated in their function. *ASIC-3* channels are activated when pH_o falls below 7, whereas *OGR1* is already half-maximally active at pH_o 7.4. **b** Dependence of plasma membrane protein function on duration of external acidosis. Some proton-gated channels desensitise in the continued presence of high extracellular proton concentration (e.g. *ASIC-3* desensitises fully within 5 s at pH 6), whereas other channels do not exhibit noticeable desensitisation (e.g. *TRPV1*) even during prolonged exposure to protons. Values for pH dependence are taken from Glitsch [8]

cells resulted in tumours that were significantly smaller when compared to tumours growing in control mouse strains. Moreover, tumours from OGR1 knock-out mice had fewer blood vessels and CD31-positive endothelial cells than those from wild type mice, suggesting reduced angiogenesis in tumours in OGR1 knock-out mice as the underlying reason for reduced tumour size in these animals. This may indicate that the acidic microenvironment of tumours promotes vessel formation in the host tissue by stimulating OGR1, thereby supporting tumour growth [17]. Similarly, knock-down of GPR4 resulted in reduced angiogenesis and tumour growth following injection of the breast cancer cell line 4T1 into GPR4 knock-out mice [43]. Intriguingly, the authors could show that GPR4 knock-down resulted in selective reduction of angiogenesis in response to vascular endothelial growth factor (VEGF), a potent stimulator of angiogenesis, but not in response to basic fibroblast growth factor, another proangiogenic factor, and that this decreased VEGF responsiveness was at least in part due to decreased expression of the VEGF receptor 2 (VEGF2R). This is particularly interesting since tumours overexpress VEGF, VEGF exerts its proangiogenic effects in tumours via activation of VEGFR2 (reviewed in [9]) and VEGF is produced by cancer cells in response to an acidic environment [34]. Hence, a picture is emerging in which extracellular acidosis promotes VEGF production in and secretion from cancer cells as well as VEGFR2 expression on vascular endothelial cells, which are the drivers for angiogenesis. VEGF then stimulates angiogenesis in tumours by activating VEGFR2, thereby allowing the tumours to expand.

Concerning impact of extracellular acidosis on immune cells, it was recently shown that an acidic microenvironment critically impacts on T lymphocyte function and produced an anergy-like (i.e. a functional inactive) state in which T lymphocytes displayed reduced cytolytic activity and cytokine expression. Crucially, proton pump inhibitors reversed this acidity-induced anergy in tumour-bearing mice, thereby increasing the therapeutic effects of active and adoptive immunotherapy [1, 5]. This is a significant finding since lack of antitumour immune cell activity is likely to contribute to tumour progression by providing a mechanism of immune escape for the cancerous cells [1], thereby promoting tumour progression.

Conclusions

One main challenge when studying the impact of extracellular pH on cell function is that we know very little about extracellular proton dynamics. Grotthuss first proposed in 1806 that protons diffuse by a proton-hopping mechanism, in which extra protons spread through the hydrogen bond network of water molecules reminiscent of electron movement in electric wires, thereby allowing for much faster diffusion of protons. However, the presence of mobile and fixed buffers for protons interferes with proton diffusion. Protons bound to mobile buffers diffuse more slowly than free protons, and the presence of fixed buffers further slows proton diffusion [15]. This reduced diffusion is exacerbated if concentration gradients are small [32]. Moreover, acid extrusion from cancerous cells can be limited by the slow reaction kinetics of the extracellular CO_2/bicarbonate pH buffer system, which enables acid extrusion from cells by generating bicarbonate and protons (the former can then be imported into cells to buffer free intracellular protons while the latter are left outside, thus contributing to extracellular acidosis). This process can be sped up by cells by expressing extracellular-facing carbonic anhydrases [14]. Understanding if and how much proton concentrations can vary locally with time will provide important insight into which membrane proteins are affected and how by external acidosis. Proton-activated proteins that do not desensitise in the prolonged presence of protons and are already active at physiological pH, such as OGR1, will continually communicate extracellular proton levels to the intracellular medium. However, a large number of proton-sensing channels or receptors have more acidic threshold levels and may exhibit pronounced desensitisation, meaning that they will only report larger changes in proton concentration and for just a short period of time. It is important to understand how signals generated by these proteins differ from messages generated by continually active proteins and how the distinct responses are integrated to form the overall reaction of a given cell. Understanding spatiotemporal patterns of extracellular protons and the proton-sensing receptors and channels expressed by the distinct cell types found in solid tumours will allow us to make predictions about proteins involved in responding to these changes. This may help identify therapeutic targets for combating selectively distinct types of cancers and hopefully lead to greatly improved treatment strategies that take advantage of the cancer's unique microenvironment and thereby minimise impact on healthy cells.

References

1. Bellone M, Calcinotto A, Filipazzi P, De Milito A, Fais S, Rivoltini L (2013) The acidity of the tumor microenvironment is a mechanism of immune escape that can be overcome by proton pump inhibitors. Oncoimmunology 2:e22058
2. Ben-Chaim Y, Tour O, Dascal N, Parnas I, Parnas H (2003) The M2 muscarinic G-protein-coupled receptor is voltage-sensitive. J Biol Chem 278:22482–22491
3. Borisevich N, Loznikova S, Sukhodola A, Halets I, Bryszewska M, Shcharbin D (2013) Acidosis, magnesium and acetylsalicylic acid: effects on thrombin. Spectrochim Acta A Mol Biomol Spectrosc 104:158–164
4. Bygrave FL, Benedetti A (1996) What is the concentration of calcium ions in the endoplasmic reticulum? Cel Calcium 19:547–551 PMID 8842522
5. Calcinotto A, Fillipazzi P, Grioni M, Iero M, De Militio A, Ricupito A, Cova A, Canese R, Jachetti E, Rossetti M, Huber V, Parmiani G, Generoso L, Santinami M, Borghi M, Fais S, Bellone M, Rivoltini L (2012) Modulation of microenvironment acidity reverses anergy in human and murine tumor-infiltrating T lymphocytes. Cancer Res 72:2746–2756
6. Chen YF, Chen YT, Chiu WT, Shen MR (2013) Remodeling of calcium signalling in tumor progression. J Biomed Sci 20:23
7. DeCoursey TE (2008) Voltage-gated proton channels. Cell Mol Life Sci 65:2554–2573
8. Glitsch M (2011) Protons and Ca^{2+}: Ionic allies in tumor progression? Physiology 26:252–265
9. Goel S, Duda DG, Xu L, Munn LL, Boucher Y, Fukumura D, Jain RK (2011) Normalization of the vasculature for treatment of cancer and other diseases. Physiol Rev 91:1071–1121

10. Hashim IA, Cornnell HH, Coelho RML, Abrahams D, Cunningham J, Lloyd M, Martinez GV, Gatenby RA, Gillies RJ (2011) Reduction of metastasis using a non-volatile buffer. Clin Exp Metastasis 28:841–849
11. Helmlinger G, Sckell A, Dellian M, Forbes NS, Jain RK (2002) Acid production in glycolysis-impaired tumors provides new insights into tumor metabolism. Clin Cancer Res 8:1284–1291
12. Holzer P (2009) Acid-sensitive ion channels and receptors. In: Canning BJ, Spina D (eds) Sensory nerves. Handbook of experimental pharmacology, vol 194. Springer, Berlin, 283–332
13. Huang WC, Swietach P, Vaughan-Jones RD, Ansorge O, Glitsch MD (2008) Extracelllar acidification elicits spatially and temporally distinct Ca^{2+} signals. Curr Biol 18:781–785
14. Hulikova A, Harris AL, Vaughan-Jones RD, Swietach P (2012) Acid-extrusion from tissue: the interplay between membrane transporters and pH buffers. Curr Pharm Des 18:1331–1337
15. Junge W, McLaughlin S (1987) The role of fixed and mobile buffers in the kinetics of proton movement. Biochim Biophys Acta 890:1–5
16. Levinthal C, Barkdull L, Jacobson P, Storjohann L, Van Wagenen BC, Stormann TM, Hammerland LG (2004) Modulation of group III metabotropic glutamate receptors by hydrogen ions. Pharmacology 83:88–94
17. Li H, Wang D, Singh LS, Berk M, Tan H, Zhao Z, Steinmetz R, Kirmani K, Wei G, Xu Y (2009) Abnormalities in osteoclastogenesis and decreased tumorigenesis in mice deficient for ovarian cancer G protein coupled receptor 1. PLoS ONE 4:e5705
18. Ludwig MG, Vanek M, Guerini D, Gasser JA, Jones CE, Junker U, Hofstetter H, Wolf RM, Seuwen K (2003) Proton-sensing G-protein-coupled receptors. Nature 425:93–98
19. Mahaut-Smith MP, Martinez-Pinna J, Gurung IS (2008) A role for membrane potential in regulating GPCRs? Trends Pharmacol Sci 29:421–429
20. Martinez-Pinna J, Gurung IS, Mahaut-Smith MP, Morales A (2010) Direct voltage control of endogenous lysophosphatidic acid G-protein-coupled receptors in *Xenopus* oocytes. J Physiol 588:1683–1693
21. Mason SD, Joyce JA (2011) Proteolytic networks in cancer. Trends Cell Biol 21:228–237
22. Mitchell P (1976) Vectorial chemistry and the molecular mechanics of chemiosmotic coupling: power transmission by proticity. Biochem Soc Trans 4:399–430
23. Moellering RE, Black KC, Krishnamurty C, Baggett BK, Stafford P, Rain M, Gatenby RA, Gillies RJ (2008) Acid treatment of melanoma cells selects for invasive phenotypes. Clin Exp Metastasis 25:411–425
24. Monteith GR, Davis FM, Roberts-Thomson SJ (2012) Calcium channels and pumps in cancer: changes and consequences. J Biol Chem 287:1666–1673
25. Morth JP, Pedersen BP, Toustrup-Jensen MS, Sorensen TL, Petersen J, Andersen JP, Vilsen B, Nissen P (2007) Crystal structure of the sodium-potassium pump. Nature 450:1043–1049
26. Neri D, Supuran CT (2011) Interfering with pH regulation in tumours as a therapeutic strategy. Nat Rev Drug Discov 10:767–777
27. Newell K, Franchi A, Pouysségur J, Tannock I (1993) Studies with glycolysis-deficient cells suggest that production of lactic acid is not the only cause of tumor acidity. Proc Natl Acad Sci U S A 90:1127–1131
28. Parent L, Supplisson S, Loo DD, Wright EM (1992) Electrogenic properties of the cloned Na+/glucose cotransporter: I. Voltage-clamp studies. J Membr Biol 125:49–62
29. Quinn SJ, Bai M, Brown EM (2004) pH sensing by the Calcium-sensing Receptor. J Biol Chem 279:37241–37249
30. Robey IF, Baggett BK, Kirkpatrick ND, Roe DJ, Dosescu J, Sloane BF, Hashim AJ, Morse DL, Raghunand N, Gatenby RA, Gillies RJ (2009) Bicarbonate increases tumor pH and inhibits spontaneous metastases. Cancer Res 69:2260–2268
31. Rofstad EK, Mathiesen B, Kindem K, Galappathi K (2006) Acidic extracellular pH promotes experimental metastasis of human melanoma cells in athymic nude mice. Cancer Res 66:6699–6707
32. Schornack PA, Gillies RJ (2003) contributions of cell metabolism and H + diffusion to the acidic pH of tumors. Neoplasia 5:135–145

33. Seuwen K, Ludwig MG, Wolf RM (2006) Receptors for protons or lipid messengers or both? J Recept Signal Transduct Res 26:599–610
34. Shi Q, Le X, Wang B, Abbruzzese JL, Xiong Q, He Y, Xie K (2001) Regulation of vascular endothelial growth factor expressin by acidosis in human cancer cells. Oncogene 20:3751–3756
35. Swietach P, Vaughan-Jones R, Harris A (2007) Regulation of tumor pH and the role of carbonic anhydrase 9. Cancer Metastasis Rev 26:299–310
36. Tominaga M, Caterina MJ, Malmberg AB, Rosen TA, Gilbert H, Skinner K, Raumann BE, Basbaum AI, Julius D (1998) The cloned capsaicin receptor integrates multiple pain-producing stimuli. Neuron 21:531–543
37. Villalba-Galea CA (2012) Voltage-controlled enzymes: the new janus bifrons. Front Pharmacol 3:161
38. Virginio C, Church D, North RA, Surprenant A (1997) Effects of divalent cations, protons and calmidazolium at the rat P2X7 receptor. Neuropharmacology 36:1285–1294
39. Waldmann R, Chamnpigny G, Bassilana F, Heurteaux C, Lazdunski M (1997) A proton-gated cation channel involved in acid-sensing. Nature 386:173–177
40. Wang Y, Li SJ, Pan J, Che Y, Yin Z, Zhao Q (2011) Specific expression of the human voltage-gated proton channel Hv1 in highly metastatic breast cancer cells, promotes tumor progression and metastasis. Biochem Biophys Res Commun 412:353–359
41. Wang Y, Li SJ, Wu X, Che Y, Li Q (2012) Clinicopathological and biological significance of human voltage-gated proton channel Hv1 protein overexpression in breast cancer. J Biol Chem 287:13877–13888
42. Webb BA, Chimenti M, Jacobson MP, Barber DL (2011) Dysregulated pH: a perfect storm for cancer progression. Nat Rev Cancer 11:671–677
43. Wyder L, Suply T, Ricoux B, Billy E, Schnell C, Baumgaren BU, Maira SM, Koelbing C, Ferretti M, Kinzel B, Müller M, Seuwen K, Ludwig MG (2011) Reduced pathological angiogenesis and tumor growth in mice lacking GPR4, a proton sensing receptor. Angiogenesis 14:533–544
44. Yamagata M, Hasuda K, Stamato T, Tannock IF (1998) The contribution of lactic acid to acidification of tumours: studies of variant cells lacking lactate dehydrogenase. Br J Cancer 77:1726–1731
45. Zheng J, Shen W, He DZZ, Long KB, Madison LD, Dallos P (2000) Prestin is the motor protein of cochlear outer hair cells. Nature 405:149–155

Chapter 8
A Genomic Analysis of Cellular Responses and Adaptions to Extracellular Acidosis

Melissa M. Keenan, Chao-Chieh Lin and Jen-Tsan Ashley Chi

The Physiological Alterations in Solid Tumor Microenvironments

The tumor microenvironment (TME) differs in multiple physiological ways from that of normal tissues. Among these physiological changes, the TME is characterized by oxygen depletion, glucose and other nutrient deprivation, high lactate levels (lactosis), and extracellular acidosis [1]. In addition to these biochemical changes, there are a number of biophysical changes within the TME, which are discussed in detail elsewhere [1]. These physiological and physical changes that define the TME are largely caused by abnormal tumor vasculature and dysregulated tumor metabolism. Many characteristics of the TME present challenges to cancer cell survival, and so are referred to as "stresses." These stresses directly or indirectly trigger tumor progression and confer treatment resistance. With great variations known to exist among different tumors, pretreatment assessment of these physiological parameters is needed to allow for the selection of appropriate therapeutic strategies for individual patients.

An accumulation of extracellular lactate and low pH, often called lactic acidosis (LA), is one physiological alteration that is found in most solid cancer tumors [2]. (The processes that cause acidification of the extracellular TME are also discussed in detail in Chap. 7). Since tumor acidosis is frequently caused by the accumulation of lactic acid, lactate levels can be used as a surrogate marker for acidosis. LA is often thought to be a simple reflection of tumor hypoxia; however, lactic acid can accumulate in the interstitial fluids of tumors because of various reasons, both as a result of, and independent from, oxygen levels. Under low oxygen, LA can be generated as a by-product of anaerobic glycolysis as cells shift to an anaerobic mode of energy production. Some tumors also exhibit a predisposition toward glycolysis even in the presence of oxygen, a phenomenon that is referred to as aerobic

J.-T. A. Chi (✉) · M. M. Keenan · C.-C. Lin
Department of Molecular Genetics and Microbiology, Center for Genomic and Computational Biology, Duke Medical Center, Durham, NC 27708, USA
e-mail: jentsan.chi@duke.edu

J-T. A. Chi (ed.), *Molecular Genetics of Dysregulated pH Homeostasis*,
DOI 10.1007/978-1-4939-1683-2_8

glycolysis or The Warburg Effect [3]. Regardless of oxygen levels, lactate and acid are excreted from cells to cause an extracellular LA. Some tumors express more proton transporters, and therefore possess a greater capacity to pump protons out into the extracellular space to create a reversed pH gradient—acidic extracellular pH (pHe) and alkaline intracellular pH (pHi) [4]. Additionally, since the perfusion and lymphatic systems of tumors are often poorly developed, LA can accumulate in tumors due to inefficient removal from the interstitial space. Therefore, LA and hypoxia in tumors are not always linked, which helps explain the disparities in the spatial and temporal distribution of hypoxia and LA in tumors [5–8]. Overall, LA is a prominent feature that influences the biology within the TME.

In tumors, the distinction between hypoxia and LA is critical for directing treatment strategies, as different therapies exist to target each of these two stresses. Treatments based on physiological manipulation (e.g., hyperthermia) may increase tumor blood perfusion to decrease both factors, whereas other approaches targeting hypoxia pathways may not relieve LA at the same time. Several recent reports have employed sodium bicarbonate to relieve tumor acidosis and reduce metastasis [9]. Tumor acidity may also change the protonation status and biological activities of chemotherapeutic agents in ways not seen with hypoxia [10]. Because the actual levels and patterns of hypoxia and LA are known to vary among cancer types, individuals, spatially and temporally, precisely identifying the severity of these two factors in each tumor will be essential to tailor individualized therapy based on these factors.

A major step toward accurately targeting these microenvironmental stresses with treatments is directly measuring them in vivo, for which a number of methods are currently in practice. Measurements of oxygen tension (via EF5, Eppendorf polarographic probe), acidity (pH probes), lactate/glucose levels (bioluminescence technique), and other microcirculatory characteristics [11–14] help to map the within-tumor distributions of metabolites reflective of variations in these stresses. Manipulating or targeting TMEs may impact clinical risk assessment, tumor behaviors, and clinical outcomes [11, 15–19]. Therefore, it is of value to identify molecular mechanisms underlying the complex and interacting individual components of in vivo TME stress responses. Unfortunately, these measurement assays are frequently invasive or require tumors to be snap-frozen in a sophisticated laboratory setting, and so are not applicable as a clinical routine. It is also difficult to analyze these stress measurements to gain molecular insights into how these differences are linked to particular oncogenic states that require differential treatment.

Studying Tumor Microenvironments In Vitro (Traditional Single-Gene Approach)

By modulating cell culture conditions, it is possible to model individual TME stresses in vitro and characterize the responses of mammalian cells. This offers a powerful means to dissect how individual environmental factors affect the behavior and phenotype of cancer cells. Various genetic and pharmacological manipulations can then be applied to evaluate the underlying molecular mechanisms and genetic circuitry. In these experiments, inferences about contributions to tumor progression rely on the observed cellular behavior. For example, a high level of lactate (lactosis) can prompt cancer cells to activate hypoxia pathways, increase CD44 expression, alter the NAD/NAD$^+$ balance, and activate CD44 and hyaluronan expression in fibroblasts [19–22]. Acidosis, on the other hand, can increase angiogenesis, cell migration, and tissue remodeling [23, 24]. Hypoxia, when applied in vitro, has been shown to promote angiogenesis, cellular migration, and energy consumption, thus providing the potential mechanisms for its association with poor clinical outcome [25, 26]. Similarly, LA, when applied to cultured cells, has been shown to trigger calcium signaling [27], pro-angiogenic gene expression (e.g., VEGF and interleukin 8, IL8) [23, 24, 28], HIF1α stabilization [29], cell death [30], and general gene expression [31–33].

An excellent study by Wojtkowiak et al. examined acid-conditioned cells by gene expression analysis [34]. They found a significant induction of ATG5 and BNIP3, genes that encode proteins involved in the autophagy pathways. The involvement of autophagy was further confirmed by an increase in the LC3-positive punctate vesicles, double-membrane vacuoles, and decreased activities of Akt and mTOR, consistent with another finding that acidosis induces a "starvation response" [35]. Interestingly, such elevated autophagy markers are maintained chronically in the acid-conditioned cells. These results argue that acidic conditions in the TME promote autophagy, and that chronic autophagy occurs as a survival adaptation under chronic acidosis.

Collectively, these studies clearly demonstrate the important roles for these stresses in tumor progression and metastasis. However, from these studies alone, the in vivo relevance of these observations for human tumors is not clear and so further understanding of the cellular changes under stress, both in vitro and in vivo, is warranted.

Transcriptomic Analysis of the Cellular Response to Acidosis (Genomic Approach)

Although tumor LA has long been recognized as an important factor, relatively little is known about how LA impacts cellular and cancer phenotypes. Many of the studies that have been performed to better understand cell response to acidic environments have done so on a single gene or in a hypothesis-driven manner as discussed above. While we learn important insights from these studies, a genomic approach to study cellular responses to acidosis can help to broaden our understanding of the full spectrum of changes and unexpected alterations elicited by LA or acidosis alone.

One of the initial genomic approaches applied to acidosis in cancer was transcriptional gene expression profiling. These genome-scale studies have been conducted to dissect the transcriptional responses of various primary non-transformed and cancer cells to LA in vitro [35, 36]. Additional analyses during these experiments compared and contrasted the transcriptional profiles from LA, acidosis, glucose deprivation, and hypoxia to gain an understanding of how cellular responses to each stress are unique or similar. A significant initial finding of these studies was that the transcriptional responses to hypoxia and LA were in fact quite different, further evidence that they have independent biological effects [36]. With Bayesian multivariate regression analysis criteria, LA induced transcriptional changes in 1585 genes, while hypoxia only changed the expression of 217; only 54 genes overlapped between the two treatments. LA treatment downregulated cell cycle, proliferation, and RNA and glucose metabolism genes, while it upregulated genes in G-protein-coupled receptor (GPCR) signaling, antigen processing and presentation, and cellular catabolism. Another study also suggested that acidosis affected cell cycle in murine breast cancer cells; under glucose deprivation, simultaneous acidosis treatment caused an arrest at the G_0/G_1 cell cycle stage and thus prevented cell death [37]. While hypoxia and LA had opposite effects on hexose and glucose metabolism, both stresses downregulated cell cycle and RNA metabolism. These initial studies clearly demonstrated dramatic differences in mammalian cells' response to hypoxia and LA.

One of these early studies also compared the transcriptional response of human mammary epithelial cells (HMECs) to acidosis and lactosis alone or together as LA. Acidosis induced much more dramatic changes in gene expression than lactosis (Fig. 8.1a). This result was also seen by the group investigating the role of LA in the inhibition of cell death to glucose deprivation; lactosis had very little effect on cell survival while acidosis inhibited cell death under glucose deprivation [37]. Acidosis induced many of the same genes as were induced by LA. The expression level of genes changed in the four groups (hypoxia, acid, lactate, lactic acid) was compared in order to assess the relative contributions of acidosis and lactosis to the LA response [36]. There was a high concordance between the LA and acidosis transcriptional responses. In contrast, this concordance was not present in the other pair-wise comparisons of lactosis versus LA, hypoxia versus LA, or hypoxia versus acidosis treatment. This suggests that LA and acidosis trigger similar genetic responses, which are distinct from the genetic responses to lactosis and hypoxia. However, the presence of lactate significantly enhanced the acidosis response to exhibit a more robust LA response.

Additional gene expression profiling studies in breast cancer cells investigated the interactions between cellular transcriptional responses to LA and hypoxia after stronger stress treatments [42]. In this study, the authors found that LA inhibited the hypoxia response by abolishing the hypoxia-induced stabilization of HIF-1α protein. Global analysis of the transcriptional changes revealed three prominent clusters of genes in the combined LA and hypoxia stress transcriptional response. The first group of genes represented LA-resistant hypoxia genes including VEGFA, HIG2, and CYP61. The second group of genes was enriched in mitogen-activated protein kinase (MAPK) and toll-like receptor signaling pathways as defined by KEGG. Within this second group, there were several inflammatory

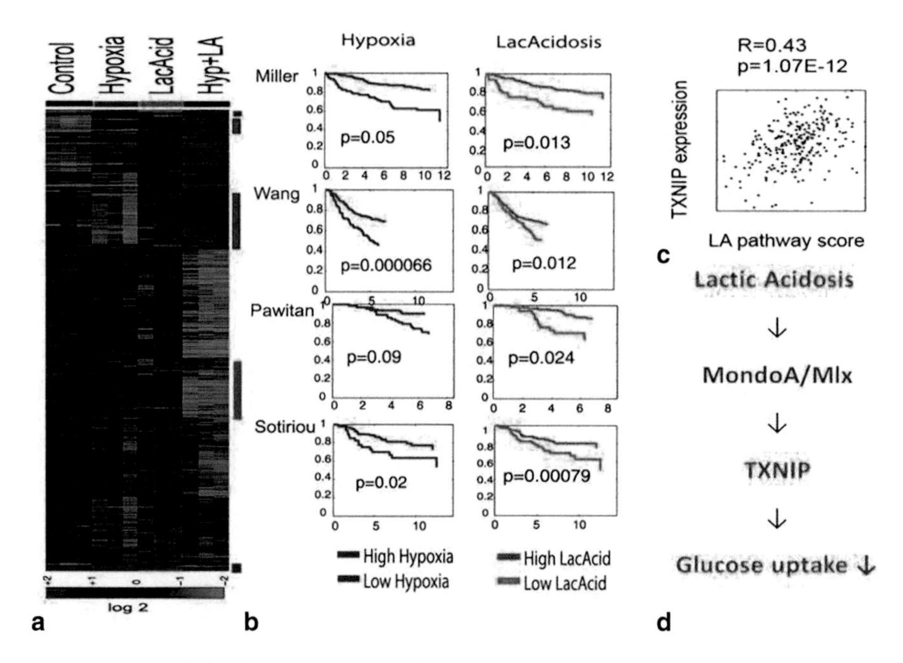

Fig. 8.1 **a** Transcriptional response of hypoxia, *lactic acidosis* (*LA*) and hypoxia + LA **b** The prognostic significance of *hypoxia* and *LA* signatures in the four indicated breast cancer datasets **c** The significant positive correlation between *TXNIP* expression and *LA pathway* in breast tumors **d** The model by which *LA* induces *TXNIP* expression through the activation of *MondoA/Mlx* complex

response genes, including TNFα, TNFAIP3, and GADD45B, which are known to be mediated by the NF-κB pathway. Interestingly, regulators of NF-κB, such as BCL3, ZFP36, and NFKBIA, were also induced by the combined treatment of LA and hypoxia. These data suggest that combined hypoxia and LA conditions lead to the activation of the NF-κB pathway. This is consistent with other transcriptomic studies that found that acidosis induced a pro-inflammatory response [39] and an inflammatory response through NF-kB [38]. The third group of genes induced by combined LA and hypoxia in Tang et al. was linked to the unfolding protein response (UPR) or ER stress pathways by CHOP, XBP-1, and ATF3 induction. Importantly, the UPR pathway shares features with the amino acid response (AAR) pathway, and both pathways seem important for cell survival in TME conditions [43, 44]. Tang et al. went on to show that the induction of the UPR pathway through ATF4 is critical for cells to respond to a combined low oxygen and LA stress, but made the significant point that different strengths and combinations of stress likely require differential stress responses. Translating this concept in vivo suggests that cells likely alter their specific stress responses constantly to the current level stress in their immediate environment.

Beyond epithelial cells, vascular endothelial cells also experience varying pH levels depending on the oxygenation status and pH of the blood, such as under acute vascular blockage. Consistent with the upregulation of GPCR signaling under LA in epithelial cells [36], one particular GPCR, GPR4, was identified to be upregulated under acidosis in vascular endothelial cells. A transcriptional response of the GPR4-dependent acidosis response revealed that, through GPR4, acidosis activated NF-kB-dependent inflammatory signals and general stress response genes and downregulated DNA-dependent transcription and nucleotide metabolism [38]. The induction of proinflammatory cytokines by acidosis was also replicated in a study of Madin-Darby canine kidney cells [39]. This study also showed similarities and differences in the cellular response to acidosis depending on the cause of the decreased pH (e.g., isocapnic or hypercapnic acidosis).

Another frequent appearance of acidosis in biology is in acid–base homeostasis achieved by the kidney during metabolic acidosis. Several studies have applied transcriptomics to study the kidney under metabolic acidosis. With somewhat less stringent cutoff criteria, one study found more than 4000 genes differentially expressed in the mouse kidney after 2 or 7 days of metabolic acidosis [33]. This and other studies of rat kidney cells under metabolic acidosis show differential gene expression in different portions of the kidney; this likely contributes to the large number of genes changed in the total kidney extracts. Additionally, many of the changes that were seen at 2 days were reverted to normal by 7 days, suggesting an in vivo mechanism of successful adaptation to acidosis. As was seen in Chen et al. 2008, the pathway most strongly affected by acidosis was oxidative phosphorylation, while the largest number of genes changed was in the solute carrier transporters functional group. Hierarchical clustering of the ~4000 genes showed six different clusters of gene expression patterns over time. The small cluster that was upregulated only at 7 days of acid loading included cytoskeleton and Wnt signaling pathways, as well as the reoccurring pattern of small GTPase and GPCR signaling. The authors also mention the high level of concordance between their mRNA abundance changes and proteomic studies of kidney cells under metabolic acidosis [33, 40]. Overall, the study confirmed that many metabolic and ion-homeostasis genes change expression under metabolic acidosis, as well as identified genes potentially involved in the chronic adaptation to metabolic acidosis.

Studies that investigate the effects of any single TME stress, such as acidosis, are critical for understanding adaptive molecular mechanisms. Yet, within solid tumors, cancer cells often experience acidosis in combination with other stresses. Some studies have begun to examine the transcriptomic effects of multiple simultaneous stresses. In the previously mentioned study in HMECs, some of the hypoxia-induced changes were repressed by the simultaneous treatment of hypoxia and LA; while a subset of LA-induced genes were further induced with the combined treatment [36]. The authors' analysis also revealed 127 and 320 genes induced or repressed, respectively, only in the combined hypoxia and LA treatment. The 127 induced genes were enriched in transcription factors, while the repressed genes were enriched in pro-apoptotic genes. It was suggested that the inhibition of apoptotic processes may be required for the cells to survive the combined stresses.

Another study further investigated the transcriptomic response to a combination of TME stresses. Specifically, this study compared the global transcriptional responses of breast cancer cells in response to three distinct TME stresses: LA, glucose deprivation, and hypoxia [35]. This study found that LA and glucose deprivation trigger highly similar transcriptional responses, each inducing features of the starvation response. However, in contrast to their comparable effects on gene expression, LA and glucose deprivation showed opposing effects on glucose uptake. This divergence of metabolic responses in the context of highly similar transcriptional responses allowed for the authors to identify a small subset of genes regulated in opposite directions by these two conditions. Among these selected genes, TXNIP and its paralogue ARRDC4 were both induced under LA and repressed with glucose deprivation. Induction of TXNIP under LA was caused by the activation of the glucose-sensing and glucose-stimulated transcriptional complex MondoA:Mlx [41]. Therefore, the upregulation of TXNIP significantly contributed to the inhibition of a glycolytic phenotype under LA [35]. Expression levels of TXNIP and ARRDC4 in human cancers were highly correlated with predicted LA pathway activities (Fig. 8.1c) and associated with favorable clinical outcomes. This integrative analysis of transcriptome and metabolic response data revealed how LA triggers features of a starvation response, while also activating the glucose-sensing MondoA-TXNIP pathway and contributing to the antitumor properties of cancer cells. These results helped open new paths to explore how these stresses influence phenotypic and metabolic adaptations in human cancers.

Collectively, these transcriptomic studies have greatly increased our knowledge and understanding of non-transformed and cancerous cells' responses to acidosis and LA. While not exhaustive across tissue, cell type, time of treatment, or strength of treatment, there are considerable parallels and similarities to mammalian cells' response to acidosis. While acidosis alone elicits significant cellular changes, it is important to remember that, in vivo, acidosis often occurs in combination with other physiological changes. Transcriptomic studies show us that dissecting the combined treatment of TME stresses is complicated, but does reveal the vast changes elicited by acidosis and LA. To better understand the importance of acidosis and LA in human cancers, we next reflect on the how these in vitro transcriptomic studies can be projected to in vivo gene expression datasets.

Projection of Hypoxia and Acidosis Gene Expression Signatures to Human Tumors

Global gene expression approaches have led to a greater understanding of acidosis and other stresses in human cancers, but it remains unclear how best to translate these gene expression changes that occur under defined cell culture manipulations in vitro to the complex behaviors of human cancers in vivo (Fig. 8.1b). Microarray-based gene expression signatures provide an approach, creating "surrogate phenotypes" of in vitro states that can be assessed in vivo. In this context, a gene signature refers to the set of genes that are most consistently and robustly changed, both up and down, by a particular treatment. Thus, these gene expression signatures from perturbations of cultured cells in vitro are used to represent a defined biological process [45–47], and these signatures can serve to recognize similar gene expression patterns in human cancer samples in vivo.

Underlying this concept is the realization that virtually any biological condition, whether a developmental state, a cellular response to extracellular ligands, or a pathological state, is reflected in changes in gene expression. While no single gene would have the full power to define the biological state, *patterns* in large-scale gene expression can reflect quite subtle distinctions in biology. Further, expression signatures are *portable:* they can be assayed in varied contexts, and so provide the capacity to link otherwise heterologous systems. A cell culture phenotype such as pathway activation is difficult to represent in a diverse sample such as a tumor. In contrast, an expression profile offers a mechanism to link two states: Expression signatures are common phenotypes shared by experimental cell culture and human tumors. For example, a hypoxia signature obtained when cultured cells are exposed to hypoxia allows the recognition of the molecular features common to multiple cancer types—in turn permitting the identification of patients with high clinical risks due to strong hypoxia response [46]. This approach has also been used to show that wound healing [48], vascular injury responses [49], various oncogenic mutations [45, 50–55], and LA [56] can play important roles in tumor progression. Additionally, linking prognostic molecular signatures of human cancers to *ex vivo* experimental cell culture models provides a relevant and controlled system that can be used in mechanistic studies and development of targeted therapeutics. Patients who are most likely to benefit from targeted therapeutics can then be recognized by high expression of the relevant gene signatures in their tumors. Substantial synergy and potential for novel biological insights can be obtained by reciprocal flow of information between the in vitro and in vivo systems.

"Top-down" approaches have been used to identify gene expression-based predictors of cancer outcomes in which no specific biological processes are associated with the tumor phenotypes [51, 57–64]. In contrast, "bottom-up" approaches define signatures based on known perturbations in cultured cells, either collected in individual experimental perturbations [45–48, 51, 65, 66] or analyzed en masse from a large collection of gene signatures, such as Gene Set Enrichment Analysis (GSEA) [67] to interrogate human cancer data. Signatures can serve as numerical factors that may improve clinical prognosis in predictive models of outcomes, including improved stratification of tumors into groups with distinct biological phenotypes. This is exemplified in a previous study of signatures reflecting responses to serum stimulation and hypoxia that led the authors to conclude that wound healing and hypoxia responses play important roles in tumor progression [46, 48]. Similar approaches have also been used to identify proliferation responses, various oncogenic pathways' deregulation, and even to predict the effectiveness of pathway-targeted therapeutics [45–47, 49, 51, 56, 68]. Gene expression signatures can also be used with functional gene ontology annotation tools to examine whether certain gene sets are enriched in particular tumor phenotypes [67].

As mentioned previously, the approach of generating "bottom-up" gene signatures in vitro, then projecting them to human tumors for similarity of gene expression responses has already been accomplished with hypoxia [69, 70], acidosis, and LA [35, 36, 42] signatures in several cancer types. Either a strong LA or acidosis alone response signature identified a subgroup of low-risk breast cancer patients with distinct metabolic profiles, suggestive of a preference for aerobic respiration. This result was consistent with the LA and acidosis signatures generated in HMECs and projected across four heterologous breast cancer datasets, as well as a LA signature generated in the MCF7 breast cancer cell line [35, 36]. When both the LA and the hypoxia signatures were used to four-way stratify breast cancer tumors based on high or low expression of either hypoxia or LA signatures, the worst prognosis was for the high hypoxia, low LA group, and better prognoses in the other three groups. The association of the LA response with good survival outcome may relate to its role in directing energy generation toward aerobic respiration and utilization of other energy sources via the inhibition of glycolysis. This "inhibition of glycolysis" phenotype in tumors is likely caused by the repression of glycolysis at the gene expression level and Akt inhibition. In fact, when smaller "glycolysis gene signatures" that reflected the changes in glycolytic genes under either hypoxia or LA were used to stratify patient outcome, they fully recapitulated the stratification seen by the entire stress gene signatures. Multiple global gene expression profiles identified the gene TXNIP as responsive to acid or LA. Additional mechanistic studies in breast cancer cells showed that a gene signature of high TXNIP activation correlated with better patent survival outcome, consistent with the LA or acid gene signatures. Clearly, the bioinformatics projection of in vitro gene signatures has significant prognostic and clinical value and further emphasizes the utility of transcriptomic studies in cancer biology.

While these sorts of projection studies have shown their relevance in oncology, they have yet to be applied to other disease states in which acidosis is a defining feature. This limitation is partially caused by the few public available datasets in other diseases, as well as the challenges of de-convoluting gene expression from mixed cell populations from benign lesions. Although with limitations, in vitro transcriptomic studies of acidosis and lactic acid have allowed us to better understand the in vivo biology through the described bioinformatics analyses [71]. To more completely understand the effect of acidosis, we cannot limit our focus to transcriptional changes; next we discuss findings from other "-omics" approaches. Since other "-omics" approaches have not been used as exhaustively in vivo, projection analyses cannot be done yet. However, these sorts of bioinformatics analyses will be critical in the future to fully integrate in vitro and in vivo data and conclusions.

Proteomic Analysis of the Cellular Response to Acidosis

Beyond transcriptomic studies, there have been a number of proteomic studies investigating mammalian cellular responses to acidosis. The vast majority of these proteomic studies have been in the context of renal adaptation to metabolic acidosis. The kidney is a major site of electrolyte exchange to maintain ion and acid–base homeostasis within the organism and so is often faced with high acid loads and acidic conditions. Under metabolic acidosis, renal cells increase glutamine uptake and catabolism, while also increasing excretion of ammonium and bicarbonate ions. To date, proteomic studies have been from a model of metabolic acidosis induced by feeding rats ammonium chloride (NH_4Cl) as their source of drinking water. These studies have looked at both acute (1 day) and chronic (7 days) effects of the metabolic acidosis treatment. In one time-course study, the authors saw that many of the changes detectable at 1 or 3 days of acidosis were corrected back to normal levels by day 7, suggesting that the kidney adapts to this stress after a period of time.

A variety of mass spectrometry techniques have been used to investigate proteomic changes due to acidosis. One concerning aspect of these proteomic studies was highlighted when two shotgun, label-free approaches were done in parallel; only 3 of 49 differentially expressed proteins after acidosis were identified by both methods [72]. The studies of rat renal cells' response to metabolic acidosis began with a general study and have since focused on particular parts of the organ, different time courses, and mitochondrial fractions of the cells. As expected, many of the proteins identified in whole-cell studies were metabolic and mitochondrial proteins. Overall, the effect of metabolic acidosis on the entire proteome was minimal, with one study only identifying 49 proteins with altered abundance after 7 days of acidosis. Such modest changes could be due to the adaptation effect, mentioned already. In particular, glutamate dehydrogenase was shown to increase expression under acidosis, which is consistent with the increase in glutamine catabolism known to occur to restore acid–base homeostasis [72]. This finding has been corroborated by other studies as well [73, 74]. Likely driven by the minimal effects seen on a cellular level, more recent studies have focused on specific portions of the proximal convoluted tubule cell. When the effect of metabolic acidosis on the apical membrane of the proximal convoluted tubule was studied over a time course, a total of 298 proteins were identified; the functionally enriched groups were consistently membrane proteins, metabolic enzymes, hydrolases, and transporters, regardless of treatment. While a number of proteins were validated to increase under acidosis (SLC5A2, DAB2, Myosin 9, SLC5A8, TMM27), they varied in their temporal dynamics and were generally not changing more than threefold [75]. Uniquely, this studied noted that more proteins decreased than increased in abundance in the acute stress of 1 or 3 days.

More recent papers have begun to investigate the proteomics of metabolic acidosis on an organelle level, focusing on the mitochondria. The first study to focus on the mitochondria found 901 proteins, 33 of which had differential abundance of >1.5-fold in the acidosis versus control rats [76]. The authors validated increased expression of five proteins (KGA, CA5B, CAT, ACAA1, HSD17B4) under acidosis and noted that a group of the 33 upregulated proteins had pH-responsive elements in their mRNAs. These pH-responsive elements are known to stabilize mRNAs under acidic conditions and so could explain the increased protein levels. There were very few proteins uniquely identified in one of the two treatments; 12 proteins were unique in the control group and 11 in acidosis. Again, as expected, many enzymes involved in glutamine catabolism increased in abundance under acidosis. Another study by Freund et al. focused their proteomic analysis on only the inner mitochondrial membrane of rat renal proximal convoluted tubule [72]. They successfully identify 206 proteins, including transmembrane proteins as expected. While not completed under acidosis, this study exemplifies the improved technical advances in the proteomic research community that will continue to improve our understanding of acidosis through future research.

An interesting trend in these datasets is the identification of acetyl-lysine residues within a small subset of the detected proteins. Acetylated-lysine residues are a recently described posttranslational modification, first identified on histones and later on multiple mitochondrial and some cytosolic proteins. While studying the mitochondrial proteome, Freund et al. found 37 acetyl-lysine residues, including 22 novel ones [72, 76]. In the inner mitochondrial membrane of rat proximal convoluted tubules, 14 proteins had N-epsilon-acetyl-lysine residues, 7 of which had not been previously identified [72]; this was approximately 6 % of the total number of identified proteins in this membrane under basal conditions. While these modified residues have been identified under acidic conditions, the biological importance or function of these posttranslational modifications remains to be determined. As proteomic techniques and reagents to study posttranslationally modified proteins continue to improve, we look forward to the advances they will bring the scientific community in understanding the importance of acidosis to the dynamic proteome.

Metabolomic Analysis of the Cellular Response to Acidosis

While both transcriptomic and proteomic experiments indicate that significant acidosis triggers metabolic reprogramming, very little is known about the metabolic flux and resulting metabolic vulnerability. While certain metabolic changes can be inferred by changes in RNA and protein levels, it is not clear whether and how these changes fully reflect the metabolic reprogramming. To formally define the metabolic reprogramming under acidosis, a recent study by LaMonte et al. used stable-isotope tracers of glucose, glutamine, and palmitate to examine how acidosis effects the central metabolic pathways of cancer cells [74] (Fig. 8.2).

Breast cancer cells exposed to acidosis have higher levels of reactive oxygen species (ROS). In order to neutralize the increase in ROS, this metabolic flux study showed that cells conserved NADPH, the reducing agent responsible for recycling GSSG (oxidized glutathione) to GSH (reduced glutathione). The need to conserve NADPH caused a number of specific changes to cellular metabolism. One change included enhanced flow through the pentose phosphate pathway (PPP) through the

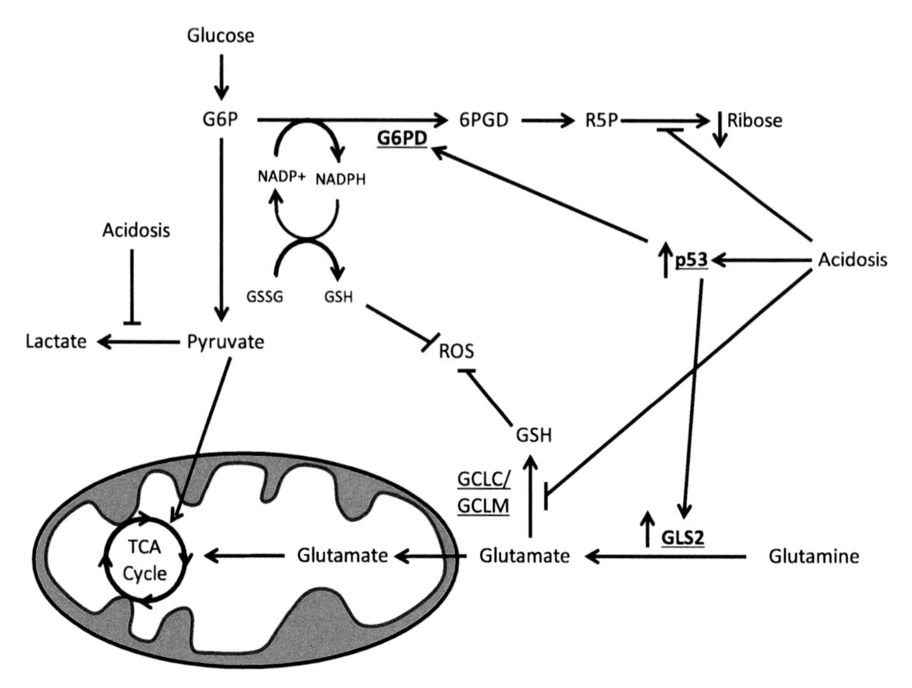

Fig. 8.2 Schematic representing the effect of acidosis on breast cancer metabolic reprogramming based on the isotope flux analysis

induction of glucose-6-phosphate dehydrogenase (*G6PD*) at the protein and mRNA levels. The oxidative branch of the PPP is the major method of NADPH synthesis in the cell. Increased shunting of glucose to the oxidative branch of the PPP generated higher levels of NADPH, and allowed cells to cope with the increased ROS. The importance of G6PD induction was shown by increased cell death under acidosis with the silencing of G6PD. Therefore, due to reduced glutathione synthesis and increased ROS present under acidosis, this stress renders breast cancer cells reliant upon cellular NADPH pools, largely from the oxidative PPP.

The redirection of glucose away from glycolysis and toward the PPP was coupled with an increased bioenergetic need for glutamine to drive the tricarboxylic acid cycle (TCA) cycle. Through the process of glutaminolysis, glutamine is metabolized to glutamate. Increased glutaminolysis was mediated by both an inhibition of glutamine synthesis, via the downregulation of glutamine synthetase (*GLUL*), and a direct upregulation of glutaminolysis, mediated by the induction of *GLS2*. Glutamate was then metabolized to α-ketoglutarate, which is able to fuel the TCA cycle, by aspartate transamination reactions (*GOT 1* and *GOT2*). Aspartate transamination reactions were strongly favored under acidosis as deamination was inhibited by the downregulation of *GLUD1,* and alanine transamination was inhibited by the downregulation of *GPT* and *GPT2*. Somewhat paradoxically, there was reduced production of glutathione (derived from glutamate) through the repression of *GCLC* and *GCLM,* which indicated that NADPH was necessary to regenerate existing GSSG due to the lack of new synthesis. As has been proposed for hypoxia, acidosis rendered these breast cancer cells increasingly reliant upon glutamine/glutamate metabolism to satisfy cellular bioenergetic demands under stress.

TP53, a major tumor suppressor, played a significant role in this metabolic reprogramming under acidosis. p53 was activated under acidosis and contributed to both the glucose and glutamine phenotype by transcriptionally inducing *G6PD* and *GLS2,* respectively. These data therefore presented a new role for p53, in which it responds to acidosis by redirecting cellular metabolism toward mitochondrial metabolism and glutaminolysis, while simultaneously acting to mitigate the ROS resulting from increased oxidative phosphorylation. Collectively, these results indicated that acidosis triggers extensive metabolic reprogramming, causing shifts in both glucose and glutamine metabolisms.

Similar metabolic experiments should be expanded to a much larger panel of cancer cell lines with known genetic makeup to understand how the different genetic mutations impact acidic metabolic reprogramming. It is likely that certain genetic mutations will allow tumor cells to survive better under acidosis and thus provide the fitness advantage necessary to clonally expand in the TME. As has been done in other disease settings, it will be important to extend these in vitro metabolomic studies in vivo in the metabolomic data of human tumors [77] to understand the metabolic similarities and differences between the systems and to use the information to inform therapeutic strategies.

Improving the Therapeutic Targeting of Acidosis Through Functional Genomic Approaches

There is a significant rationale for the development of agents that target tumor cells specifically under LA. Although short-term transcriptional analysis of LA revealed the inhibition of growth and glycolysis of primary tumors by LA, there is considerable evidence to suggest that LA is a critical adverse factor for overall clinical outcomes as it can: (1) increase metastatic potential through increased mobility/migration and activation of proteolytic enzymes (MMP-2, MMP-9) [78, 79]; (2) confer resistance to tumor therapeutics due to the drug protonation status or an increase in the activities of multi-drug resistance 1 (*MDR1*) [80, 81]; and (3) select for cancer cells with more invasive and stem cell-like phenotypes [31, 82–84]. It is important to note that these effects of acidosis are on the mobility/migration and chemoresistance abilities of tumor cells, which are distinct from the anti-growth and suppression of proliferation by acidosis. Therefore, it would be of great benefit to relieve LA to selectively eliminate the cells that exhibit these aggressive growth behaviors and/or resist standard interventions. To reduce tumor acidity in the clinic, systemic bicarbonate buffers have been used with some biological effects [9, 85], but the general applicability and long-term consequences of overloading patients with systemic bicarbonate buffers remain to be determined. Thus, there exists a significant and largely unmet need to eradicate cancer cells associated with LA [2].

In classic genetics, two genes are deemed "synthetically lethal" if mutations in either gene alone are compatible with viability, but simultaneous mutation of both genes leads to death [86]. Recently, this concept has been extended to include genes that are contextually essential: for example, critical for cell survival under stress (hypoxia or LA), but not under normal conditions. Such a strategy relies on the unique features of solid tumors' stresses, which are not found in nonmalignant tissues under tightly controlled pO_2 and pH. Therapeutic strategies targeting these stresses would be expected to have a higher specificity and therapeutic index.

Since both hypoxia and LA are tumor-specific features, and, unlike some oncogenic mutations, are not easily directly inhibited pharmacologically, it is not immediately obvious how best to eradicate these cells. Synthetic lethality can be a useful concept to exploit in the development of strategies to target these cells [87]. Current therapeutic strategies are mainly based on the known putative survival mechanisms instead of unbiased genetic identification. For example, the inhibition of glucose uptake, inhibition of the expression and activity of the HIF proteins, or critical steps in the *ATF4*-driven UPR have been proposed as methods to target cells under hypoxia [88]. A recent report also suggests a synthetically lethal relationship between PARP1 inhibition and hypoxia [89]. While there are currently limited options for targeting hypoxia, there are even fewer targeting cells under acidosis or LA. Most of the ongoing efforts in this regard have been focused on compromising the neutralization capacity of transporters and enzymes, such as the monocarboxylate transporters (MCTs) [90], the sodium–proton exchanger NHE1 [4], the carbonic anhydrases (CA9, CA12) [91], and the vacuolar proton ATPase (V-ATPase) [92]. These "proton exchangers" maintain intracellular pH under extracellular acidosis. While these efforts have led to some encouraging results, these strategies are still early in development and there is significant redundancy among different transporters for proton export. Consequently, there is a strong need to identify other therapeutic targets to target cancer cells under conditions of both hypoxia and LA.

Functional genetic screens provide an unbiased and powerful means of identifying genes responsible for any phenotype that can be measured experimentally. In recent years, the generation and development of shRNA libraries targeting the entire human, mouse, and rat genomes has greatly advanced the ability of individual investigators to deploy high-throughput genetic screens in mammalian cells [93–96]. For example, several groups independently performed RNA interference (RNAi) screens to identify genes synthetically lethal for *KRAS* mutations, a prevalent, but currently un-druggable, oncogenic mutation in cancers [97, 98]. These screens uncovered multiple points of vulnerability in *KRAS* mutations that can be exploited for therapeutic purposes. Recent studies have begun performing RNAi screens using xenografts in vivo to identify genes relevant for in vivo oncogenesis [99–101]. The improvement of therapeutics from these in vivo screens remains to be seen.

Global RNAi screens have been used in model organisms for longer than mammalian cells and so more detailed screens, such as investigating responses to specific stresses, have been conducted in this context. For example, an RNAi screen investigating the hypoxia response in *Drosophila* identified critical roles for the ATF4 homologue (*Cryptocephal*) and *Dicer1* [102]. Many other studies have been performed along these lines; we suggest the reader seek out those most relevant to his/her area of interest, as they are not the focus of this chapter. However, it is clear from model organism studies that these powerful approaches can reveal genes that exhibit synthetic lethality under stress, yet few such studies have been performed in mammalian cells.

There are numerous examples of RNAi screens, both in model organisms and mammalian systems, which suggest that the processes known to be influenced by acidosis and LA can be investigated by this experimental technique. An RNAi screen in Drosophila investigated the genes required for phosphate response, thus establishing a system for studying biological effects elicited by an ionic molecule [103]. Another group utilized Drosophila cells and an RNAi screen with an H148Q-YFP anion-sensitive indicator to understand chloride ion homeostasis mechanisms [104]. Cell–cell adhesion is relevant to the effect acidosis has on the migration and invasion of cancer cells; the conserved genes necessary for cadherin-mediated cell–cell adhesion has been investigated through an RNAi screen [105]. Interestingly, the 17 "regulatory protein hubs" mediating adhesion included GPCR signaling, metabolic processes and channels, and receptors, all of which are relevant to mammalian cells under acidic conditions based on transcriptomic analyses [36].

In the context of cancer biology, these functional RNAi screens have mostly focused on the genes necessary for the general proliferation of various cancer cell lines. This is the major effort of the Achilles' heel Project at The Broad Institute [106], which has now accomplished proliferation-RNAi screens in hundreds of cancer cell lines. A major part of this project will be to understand the molecular mechanisms underlying the genetic dependencies so that the specificity of drugs can be determined. To begin to understand context-dependent lethality, some RNAi studies have investigated the revived importance of cancer metabolism and the mitochondria to tumorigenesis. A recent study by Birsoy and colleagues created a unique experimental setup with a Nutrostat system for continuous flow of media to maintain cells under low glucose growth conditions—another TME-relevant stress—and then conducted an RNAi screen under these growth conditions [107]. While they did not look at the effect of acidosis or LA on their system, this study revealed the importance of mitochondrial metabolism to cancer cells' susceptibility to drugs and so is highly relevant to understanding the effects of acidosis [107].

Demonstrating the likely relevance of acidosis in many oncogenic processes, metabolic genes related to acidosis metabolism are enriched in or top hits from many mammalian RNAi screens. These RNAi screens, in multiple different contexts, investigated a wide range of topics: growth of brain cancer stem cells [108], reliance on IL-3 transformation [109], KRAS-dependent lung adenocarcinoma growth [110], and ccRCC tumorigenesis drivers [111]. As expected, in many of these examples, inhibiting the processes known to be activated by acidosis caused increased cancer cell death.

And yet, to our knowledge, there are no examples of an RNAi screen under LA or acidosis in mammalian cells. A potential experimental setup is illustrated in Fig. 8.3. This remains an important and therapeutically relevant line of investigation

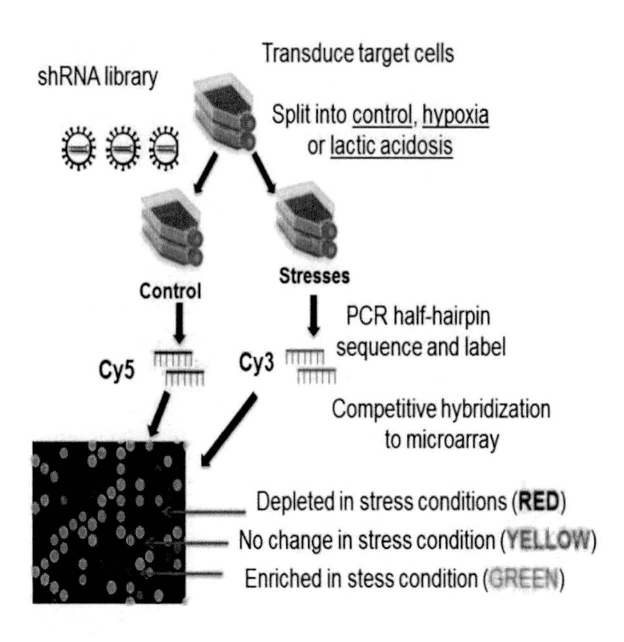

Fig. 8.3 Overview of the synthetic lethal genetic screens to identify genes/kinases which are essential for survival (*shRNA* depleted) or restrict cell growth (*shRNA* enriched) under the stresses of hypoxia or lactic acidosis

with significant potential to identify methods to target particularly problematic regions of tumors. The closest study to investigating cancer cell survival under acidosis is a recent study that performed a kinome-wide RNAi screen under anoxia [112]. This study focused on the effect that the kinome on hypoxia stress responses such as nuclear size, ER, and DNA damage stress responses and spheroid formation; they identified five kinases with novel roles in these processes (DYRK1B, GAK, IHPK2, IRAK4, and MATK). There was also an inverse relationship between NF-kB and viable cell number in hypoxia; the authors comment that there are many therapeutic strategies in progress to target NF-kB as an immune modulator. This study represents an important example of the progress toward a better understanding of contextually essential genes that can be targeted for improved therapeutics in cancer and other disease contexts.

Ongoing challenges with the RNAi-based methods of these functional genomic screens include incomplete gene inactivation and off-target effects. Therefore, various gene editing approaches, including zinc finger nucleases (ZFNs) [113] or transcription activator-like effector nucleases (TALENs) [114] have been utilized to delete or modify a genetic locus and achieve complete silencing. Recently, various gene editing methods based on the clustered, regularly interspaced, short palindromic repeats (CRISPR)-Cas9 system have gained enormous popularity due to the ease of implementation and higher efficiency compared to other genome editing techniques [115–117]. Three recent papers have applied the CRISPR systems on a large scale to perform functional genomic screens in human [118, 119] and mouse [120] cells. While these methods are in the early stages of adoption, preliminary data indicate that these methods may provide significant advantages over the RNAi-based screens due to a more complete removal of the target genes and increased specificity. These new formats for genetic manipulations will likely improve the efficiency and specificity of functional genomic screens during both the initial screening and the subsequent validation stages of these studies.

Conclusions and Future Perspectives

Many solid tumors grow quickly and expand beyond the capacity of the local blood supply, leading to regions of hypoxia and LA. Most tumors contain regions with varying levels of hypoxia and LA, which can make tumors more resistant to treatment and ultimately result in treatment failure and relapse. In addition, cancer cells experiencing hypoxia and LA tend to migrate and metastasize distantly. Even though the negative consequences of hypoxia and LA in tumors have long been recognized, there are few therapies targeting these stresses. Transcriptomic, proteomic, metabolomic, and genomic analyses of how cancer cells as well as renal tubular cells respond to acidosis and LA have revealed some crucial adaptation mechanisms that can be targeted to eradicate cancer cells under these stresses. Applying the concept of synthetic lethality, functional genomic approaches are likely to reveal additional genes and pathways which can be employed to therapeutically target tumor cells under LA and relieve patients of the clinical issues resulting from this stress in most solid tumors.

Acknowledgment Supported by NIH CA125618, CA106520, F31 CA180610 and the Department of Defense W81XWH-12-1-0148 and W81XWH-14-1-0309.

References

1. Vaupel P (2004) Tumor microenvironmental physiology and its implications for radiation oncology. Semin Radiat Oncol 14(3):198–206
2. Webb BA, Chimenti M, Jacobson MP, Barber DL (2011) Dysregulated pH: a perfect storm for cancer progression. Nat Rev Cancer 11(9):671–677
3. Gatenby RA, Gillies RJ (2004) Why do cancers have high aerobic glycolysis? Nat Rev Cancer 4(11):891–899
4. Cardone RA, Casavola V, Reshkin SJ (2005) The role of disturbed pH dynamics and the Na^+/H^+ exchanger in metastasis. Nat Rev Cancer 5(10):786–795
5. Gulledge CJ, Dewhirst MW (1996) Tumor oxygenation: a matter of supply and demand. Anticancer Res 16(2):741–749
6. Helmlinger G, Yuan F, Dellian M, Jain RK (1997) Interstitial pH and pO2 gradients in solid tumors in vivo: high-resolution measurements reveal a lack of correlation. Nat Med 3(2):177–182
7. Schornack PA, Gillies RJ (2003) Contributions of cell metabolism and H^+ diffusion to the acidic pH of tumors. Neoplasia 5(2):135–145

8. Vaupel P, Hockel M (2000) Blood supply, oxygenation status and metabolic micromilieu of breast cancers: characterization and therapeutic relevance. Int J Oncol 17(5):869–879
9. Robey IF, Baggett BK, Kirkpatrick ND, Roe DJ, Dosescu J, Sloane BF, Hashim AI, Morse DL, Raghunand N, Gatenby RA et al (2009) Bicarbonate increases tumor pH and inhibits spontaneous metastases. Cancer Res 69(6):2260–2268
10. Adams DJ (2005) The impact of tumor physiology on camptothecin-based drug development. Curr Med Chem Anticancer Agents 5(1):1–13
11. Mueller-Klieser W, Walenta S (1993) Geographical mapping of metabolites in biological tissue with quantitative bioluminescence and single photon imaging. Histochem J 25(6):407–420
12. Thews O, Kelleher DK, Vaupel PW (1995) Modulation of spatial O_2 tension distribution in experimental tumors by increasing arterial O_2 supply. Acta Oncol 34(3):291–295
13. Kallinowski F, Schlenger KH, Runkel S, Kloes M, Stohrer M, Okunieff P, Vaupel P (1989) Blood flow, metabolism, cellular microenvironment, and growth rate of human tumor xenografts. Cancer Res 49(14):3759–3764
14. Dewhirst MW, Klitzman B, Braun RD, Brizel DM, Haroon ZA, Secomb TW (2000) Review of methods used to study oxygen transport at the microcirculatory level. Int J Cancer 90(5):237–255
15. Brizel DM, Schroeder T, Scher RL, Walenta S, Clough RW, Dewhirst MW, Mueller-Klieser W (2001) Elevated tumor lactate concentrations predict for an increased risk of metastases in head-and-neck cancer. Int J Radiat Oncol Biol Phys 51(2):349–353
16. Schwickert G, Walenta S, Sundfor K, Rofstad EK, Mueller-Klieser W (1995) Correlation of high lactate levels in human cervical cancer with incidence of metastasis. Cancer Res 55(21):4757–4759
17. Walenta S, Salameh A, Lyng H, Evensen JF, Mitze M, Rofstad EK, Mueller-Klieser W (1997) Correlation of high lactate levels in head and neck tumors with incidence of metastasis. Am J Pathol 150(2):409–415
18. Walenta S, Wetterling M, Lehrke M, Schwickert G, Sundfor K, Rofstad EK, Mueller-Klieser W (2000) High lactate levels predict likelihood of metastases, tumor recurrence, and restricted patient survival in human cervical cancers. Cancer Res 60(4):916–921
19. Walenta S, Mueller-Klieser WF (2004) Lactate: mirror and motor of tumor malignancy. Semin Radiat Oncol 14(3):267–274
20. Formby B, Stern R (2003) Lactate-sensitive response elements in genes involved in hyaluronan catabolism. Biochem Biophys Res Commun 305(1):203–208
21. Stern R, Shuster S, Neudecker BA, Formby B (2002) Lactate stimulates fibroblast expression of hyaluronan and CD44: the Warburg effect revisited. Exp Cell Res 276(1):24–31
22. Lu H, Forbes RA, Verma A (2002) Hypoxia-inducible factor 1 activation by aerobic glycolysis implicates the Warburg effect in carcinogenesis. J Biol Chem 277(26):23111–23115
23. Fukumura D, Xu L, Chen Y, Gohongi T, Seed B, Jain RK (2001) Hypoxia and acidosis independently up-regulate vascular endothelial growth factor transcription in brain tumors in vivo. Cancer Res 61(16):6020–6024
24. Xu L, Fidler IJ (2000) Acidic pH-induced elevation in interleukin 8 expression by human ovarian carcinoma cells. Cancer Res 60(16):4610–4616
25. Semenza GL (2002) HIF-1 and tumor progression: pathophysiology and therapeutics. Trends Mol Med 8(Suppl 4):S62–S67
26. Harris AL (2002) Hypoxia–a key regulatory factor in tumor growth. Nat Rev Cancer 2(1):38–47
27. Huang WC, Swietach P, Vaughan-Jones RD, Ansorge O, Glitsch MD (2008) Extracellular acidification elicits spatially and temporally distinct Ca^{2+} signals. Curr Biol 18(10):781–785
28. Shi Q, Le X, Wang B, Abbruzzese JL, Xiong Q, He Y, Xie K (2001) Regulation of vascular endothelial growth factor expression by acidosis in human cancer cells. Oncogene 20(28):3751–3756
29. Mekhail K, Gunaratnam L, Bonicalzi ME, Lee S (2004) HIF activation by pH-dependent nucleolar sequestration of VHL. Nat Cell Biol 6(7):642–647

30. Graham RM, Frazier DP, Thompson JW, Haliko S, Li H, Wasserlauf BJ, Spiga MG, Bishopric NH, Webster KA (2004) A unique pathway of cardiac myocyte death caused by hypoxia-acidosis. J Exp Biol 207(Pt 18):3189–3200
31. Moellering RE, Black KC, Krishnamurty C, Baggett BK, Stafford P, Rain M, Gatenby RA, Gillies RJ (2008) Acid treatment of melanoma cells selects for invasive phenotypes. Clin Exp Metastasis 25(4):411–425
32. Zieker D, Schafer R, Glatzle J, Nieselt K, Coerper S, Northoff H, Konigsrainer A, Hunt TK, Beckert S (2008) Lactate modulates gene expression in human mesenchymal stem cells. Langenbecks Arch Surg 393(3):297–301
33. Nowik M, Lecca MR, Velic A, Rehrauer H, Brandli AW, Wagner CA (2008) Genome-wide gene expression profiling reveals renal genes regulated during metabolic acidosis. Physiol Genomics 32(3):322–334
34. Wojtkowiak JW, Rothberg JM, Kumar V, Schramm KJ, Haller E, Proemsey JB, Lloyd MC, Sloane BF, Gillies RJ (2012) Chronic autophagy is a cellular adaptation to tumor acidic pH microenvironments. Cancer Res 72(16):3938–3947
35. Chen JL, Merl D, Peterson CW, Wu J, Liu PY, Yin H, Muoio DM, Ayer DE, West M, Chi JT (2010) Lactic acidosis triggers starvation response with paradoxical induction of TXNIP through MondoA. PLoS Genet 6(9):e1001093
36. Chen JL, Lucas JE, Schroeder T, Mori S, Wu J, Nevins J, Dewhirst M, West M, Chi JT (2008) The genomic analysis of lactic acidosis and acidosis response in human cancers. PLoS Genet 4(12):e1000293
37. Wu H, Ding Z, Hu D, Sun F, Dai C, Xie J, Hu X (2012) Central role of lactic acidosis in cancer cell resistance to glucose deprivation-induced cell death. J Pathol 227(2):189–199
38. Dong L, Li Z, Leffler NR, Asch AS, Chi JT, Yang LV (2013) Acidosis activation of the proton-sensing GPR4 receptor stimulates vascular endothelial cell inflammatory responses revealed by transcriptome analysis. PLoS One 8(4):e61991
39. Raj S, Scott DR, Nguyen T, Sachs G, Kraut JA (2013) Acid stress increases gene expression of proinflammatory cytokines in Madin-Darby canine kidney cells. Am J Physiol Renal Physiol 304(1):F41–F48
40. Curthoys NP, Taylor L, Hoffert JD, Knepper MA (2007) Proteomic analysis of the adaptive response of rat renal proximal tubules to metabolic acidosis. Am J Physiol Renal Physiol 292(1):F140–F147
41. Stoltzman CA, Peterson CW, Breen KT, Muoio DM, Billin AN, Ayer DE (2008) Glucose sensing by MondoA:Mlx complexes: a role for hexokinases and direct regulation of thioredoxin-interacting protein expression. Proc Natl Acad Sci U S A 105(19):6912–6917
42. Tang X, Lucas JE, Chen JL, LaMonte G, Wu J, Wang MC, Koumenis C, Chi JT (2012) Functional interaction between responses to lactic acidosis and hypoxia regulates genomic transcriptional outputs. Cancer Res 72(2):491–502
43. Romero-Ramirez L, Cao H, Nelson D, Hammond E, Lee AH, Yoshida H, Mori K, Glimcher LH, Denko NC, Giaccia AJ et al (2004) XBP1 is essential for survival under hypoxic conditions and is required for tumor growth. Cancer Res 64(17):5943–5947
44. Rouschop KM, van den Beucken T, Dubois L, Niessen H, Bussink J, Savelkouls K, Keulers T, Mujcic H, Landuyt W, Voncken JW et al (2010) The unfolded protein response protects human tumor cells during hypoxia through regulation of the autophagy genes MAP1LC3B and ATG5. J Clin Invest 120(1):127–141
45. Bild AH, Yao G, Chang JT, Wang Q, Potti A, Chasse D, Joshi MB, Harpole D, Lancaster JM, Berchuck A et al (2005) Oncogenic pathway signatures in human cancers as a guide to targeted therapies. Nature 439:353–357
46. Chi JT, Wang Z, Nuyten DS, Rodriguez EH, Schaner ME, Salim A, Wang Y, Kristensen GB, Helland A, Borresen-Dale AL et al (2006) Gene expression programs in response to hypoxia: cell type specificity and prognostic significance in human cancers. PLoS Med 3(3):e47
47. Lamb J, Ramaswamy S, Ford HL, Contreras B, Martinez RV, Kittrell FS, Zahnow CA, Patterson N, Golub TR, Ewen ME (2003) A mechanism of cyclin D1 action encoded in the patterns of gene expression in human cancer. Cell 114(3):323–334

48. Chang HY, Sneddon JB, Alizadeh AA, Sood R, West RB, Montgomery K, Chi JT, Rijn Mv M, Botstein D, Brown PO (2004) Gene expression signature of fibroblast serum response predicts human cancer progression: similarities between tumors and wounds. PLoS Biol 2(2):E7
49. Chi J-T, Rodriguez EH, Wang Z, Nuyten DSA, Mukherjee S, de Rijn Mv, de Vijver MJv, Hastie T, Brown PO (2007) Gene expression programs of human smooth muscle cells: tissue-specific differentiation and prognostic significance in breast cancers. PLoS Genet 3(9):e164
50. Chang JT, Carvalho C, Mori S, Bild A, Gatza M, Wang Q, Lucase JE, Potti A, Febbo P, West M et al (2009) A genomic strategy to elucidate modules of oncogenic pathway signaling networks. Mol Cell 34:104–114 (Accepted)
51. Huang E, Ishida S, Pittman J, Dressman H, Bild A, Kloos M, D'Amico M, Pestell RG, West M, Nevins JR (2003) Gene expression phenotypic models that predict the activity of oncogenic pathways. Nat Genet 34(2):226–230
52. Mori S, Rempel RE, Chang JT, Yao G, Lagoo AS, Potti A, Bild A, Nevins JR (2008) Utilization of pathway signatures to reveal distinct types of B lymphoma in the Emicro-myc model and human diffuse large B-cell lymphoma. Cancer Res 68(20):8525–8534
53. Nevins JR, Potti A (2007) Mining gene expression profiles: expression signatures as cancer phenotypes. Nat Rev Genet 8(8):601–609
54. West M, Blanchette C, Dressman H, Huang E, Ishida S, Spang R, Zuzan H, Olson JA Jr, Marks JR, Nevins JR (2001) Predicting the clinical status of human breast cancer by using gene expression profiles. Proc Natl Acad Sci U S A 98(20):11462–11467
55. West M, Ginsburg GS, Huang AT, Nevins JR (2006) Embracing the complexity of genomic data for personalized medicine. Genome Res 16(5):559–566
56. Chen JL, Lucase JE, Schroeder T, Mori S, Nevins JR, Dewhirst MW, West M, Chi JT (2008) Genomic analysis of response to lactic acidosis and acidosis in human cancers. PLoS Genet 4(12):e1000293
57. Alizadeh AA, Eisen MB, Davis RE, Ma C, Lossos IS, Rosenwald A, Boldrick JC, Sabet H, Tran T, Yu X et al (2000) Distinct types of diffuse large B-cell lymphoma identified by gene expression profiling. Nature 403(6769):503–511
58. Huang E, Cheng SH, Dressman H, Pittman J, Tsou MH, Horng CF, Bild A, Iversen ES, Liao M, Chen CM et al (2003) Gene expression predictors of breast cancer outcomes. Lancet 361(9369):1590–1596
59. Perou CM, Sorlie T, Eisen MB, van de Rijn M, Jeffrey SS, Rees CA, Pollack JR, Ross DT, Johnsen H, Akslen LA et al (2000) Molecular portraits of human breast tumours. Nature 406(6797):747–752
60. Golub TR (2001) Genome-wide views of cancer. N Engl J Med 344(8):601–602
61. Golub TR (2004) Toward a functional taxonomy of cancer. Cancer Cell 6(2):107–108
62. Golub TR, Slonim DK, Tamayo P, Huard C, Gaasenbeek M, Mesirov JP, Coller H, Loh ML, Downing JR, Caligiuri MA et al (1999) Molecular classification of cancer: class discovery and class prediction by gene expression monitoring. Science 286(5439):531–537
63. van de Vijver MJ, He YD, van't Veer LJ, Dai H, Hart AA, Voskuil DW, Schreiber GJ, Peterse JL, Roberts C, Marton MJ et al (2002) A gene-expression signature as a predictor of survival in breast cancer. N Engl J Med 347(25):1999–2009
64. van't Veer LJ, Dai H, van de Vijver MJ, He YD, Hart AA, Mao M, Peterse HL, van der Kooy K, Marton MJ, Witteveen AT et al (2002) Gene expression profiling predicts clinical outcome of breast cancer. Nature 415(6871):530–536
65. Sweet-Cordero A, Mukherjee S, Subramanian A, You H, Roix JJ, Ladd-Acosta C, Mesirov J, Golub TR, Jacks T (2005) An oncogenic KRAS2 expression signature identified by cross-species gene-expression analysis. Nat Genet 37(1):48–55
66. Bild A, Febbo PG (2005) Application of a priori established gene sets to discover biologically important differential expression in microarray data. Proc Natl Acad Sci U S A 102(43):15278–15279
67. Subramanian A, Tamayo P, Mootha VK, Mukherjee S, Ebert BL, Gillette MA, Paulovich A, Pomeroy SL, Golub TR, Lander ES et al (2005) Gene set enrichment analysis: a knowledge-

based approach for interpreting genome-wide expression profiles. Proc Natl Acad Sci U S A 102(43):15545–15550
68. Whitfield ML, Sherlock G, Saldanha AJ, Murray JI, Ball CA, Alexander KE, Matese JC, Perou CM, Hurt MM, Brown PO et al (2002) Identification of genes periodically expressed in the human cell cycle and their expression in tumors. Mol Biol Cell 13(6):1977–2000
69. Winter SC, Buffa FM, Silva P, Miller C, Valentine HR, Turley H, Shah KA, Cox GJ, Corbridge RJ, Homer JJ et al (2007) Relation of a hypoxia metagene derived from head and neck cancer to prognosis of multiple cancers. Cancer Res 67(7):3441–3449
70. Chi JT, Wang Z, Nuyten DS, Rodriguez EH, Schaner ME, Salim A, Wang Y, Kristensen GB, Helland A, Borresen-Dale AL et al (2006) Gene expression programs in response to hypoxia: cell type specificity and prognostic significance in human cancers. PLoS Med 3(3):e47
71. Lucas JE, Kung HN, Chi JT (2010) Latent factor analysis to discover pathway-associated putative segmental aneuploidies in human cancers. PLoS Comput Biol 6(9):e1000920
72. Freund DM, Prenni JE, Curthoys NP (2013) Proteomic profiling of the mitochondrial inner membrane of rat renal proximal convoluted tubules. Proteomics 13(16):2495–2499
73. Zaganas I, Spanaki C, Plaitakis A (2012) Expression of human GLUD2 glutamate dehydrogenase in human tissues: functional implications. Neurochem Int 61(4):455–462
74. Lamonte G, Tang X, Chen JL, Wu J, Ding CKC, Keenan MM, Sangokoya C, Kung HN, Ilkayeva O, Boros LG et al (2014) Acidosis induces reprogramming of cellular metabolism to mitigate oxidative stress. Cancer Metab 1:23
75. Walmsley SJ, Freund DM, Curthoys NP (2012) Proteomic profiling of the effect of metabolic acidosis on the apical membrane of the proximal convoluted tubule. Am J Physiol Renal Physiol 302(11):F1465–F1477
76. Freund DM, Prenni JE, Curthoys NP (2013) Response of the mitochondrial proteome of rat renal proximal convoluted tubules to chronic metabolic acidosis. Am J Physiol Renal Physiol 304(2):F145–F155
77. Tang X, Lin CC, Spasojevic I, Iversen ES, Chi JT, Marks JR (2014) A joint analysis of metabolomics and genetics of breast cancer. Breast Cancer Res 16 (4):415
78. Rofstad EK, Mathiesen B, Kindem K, Galappathi K (2006) Acidic extracellular pH promotes experimental metastasis of human melanoma cells in athymic nude mice. Cancer Res 66(13):6699–6707
79. Martinez-Zaguilan R, Seftor EA, Seftor RE, Chu YW, Gillies RJ, Hendrix MJ (1996) Acidic pH enhances the invasive behavior of human melanoma cells. Clin Exp Metastasis 14(2):176–186
80. Sauvant C, Nowak M, Wirth C, Schneider B, Riemann A, Gekle M, Thews O (2008) Acidosis induces multi-drug resistance in rat prostate cancer cells (AT1) in vitro and in vivo by increasing the activity of the p-glycoprotein via activation of p38. Int J Cancer 123(11):2532–2542
81. De Milito A, Canese R, Marino ML, Borghi M, Iero M, Villa A, Venturi G, Lozupone F, Iessi E, Logozzi M et al (2010) pH-dependent antitumor activity of proton pump inhibitors against human melanoma is mediated by inhibition of tumor acidity. Int J Cancer 127(1):207–219
82. Gatenby RA, Gillies RJ (2008) A microenvironmental model of carcinogenesis. Nat Rev Cancer 8(1):56–61
83. Fang JS, Gillies RD, Gatenby RA (2008) Adaptation to hypoxia and acidosis in carcinogenesis and tumor progression. Semin Cancer Biol 18(5):330–337
84. Hjelmeland AB, Wu Q, Heddleston JM, Choudhary GS, MacSwords J, Lathia JD, McLendon R, Lindner D, Sloan A, Rich JN (2011) Acidic stress promotes a glioma stem cell phenotype. Cell Death Differ 18(5):829–840
85. Silva AS, Yunes JA, Gillies RJ, Gatenby RA (2009) The potential role of systemic buffers in reducing intratumoral extracellular pH and acid-mediated invasion. Cancer Res 69(6):2677–2684
86. Kaelin WG Jr (2005) The concept of synthetic lethality in the context of anticancer therapy. Nat Rev Cancer 5(9):689–698

87. Chan DA, Giaccia AJ (2011) Harnessing synthetic lethal interactions in anticancer drug discovery. Nat Rev Drug Discov 10(5):351–364
88. Wilson WR, Hay MP (2011) Targeting hypoxia in cancer therapy. Nat Rev Cancer 11(6):393–410
89. Chan N, Pires IM, Bencokova Z, Coackley C, Luoto KR, Bhogal N, Lakshman M, Gottipati P, Oliver FJ, Helleday T et al (2010) Contextual synthetic lethality of cancer cell kill based on the tumor microenvironment. Cancer Res 70(20):8045–8054
90. Sonveaux P, Vegran F, Schroeder T, Wergin MC, Verrax J, Rabbani ZN, De Saedeleer CJ, Kennedy KM, Diepart C, Jordan BF et al (2008) Targeting lactate-fueled respiration selectively kills hypoxic tumor cells in mice. J Clin Invest 118(12):3930–3942
91. Parks SK, Chiche J, Pouyssegur J (2011) pH control mechanisms of tumor survival and growth. J Cell Physiol 226(2):299–308
92. Fais S, De Milito A, You H, Qin W (2007) Targeting vacuolar H+ -ATPases as a new strategy against cancer. Cancer Res 67(22):10627–10630
93. Berns K, Hijmans EM, Mullenders J, Brummelkamp TR, Velds A, Heimerikx M, Kerkhoven RM, Madiredjo M, Nijkamp W, Weigelt B et al (2004) A large-scale RNAi screen in human cells identifies new components of the p53 pathway. Nature 428(6981):431–437
94. Luo B, Cheung HW, Subramanian A, Sharifnia T, Okamoto M, Yang X, Hinkle G, Boehm JS, Beroukhim R, Weir BA et al (2008) Highly parallel identification of essential genes in cancer cells. Proc Natl Acad Sci U S A 105(51):20380–20385
95. Schlabach MR, Luo J, Solimini NL, Hu G, Xu Q, Li MZ, Zhao Z, Smogorzewska A, Sowa ME, Ang XL et al (2008) Cancer proliferation gene discovery through functional genomics. Science 319(5863):620–624
96. Silva JM, Marran K, Parker JS, Silva J, Golding M, Schlabach MR, Elledge SJ, Hannon GJ, Chang K (2008) Profiling essential genes in human mammary cells by multiplex RNAi screening. Science 319(5863):617–620
97. Luo J, Emanuele MJ, Li D, Creighton CJ, Schlabach MR, Westbrook TF, Wong KK, Elledge SJ (2009) A genome-wide RNAi screen identifies multiple synthetic lethal interactions with the Ras oncogene. Cell 137(5):835–848
98. Scholl C, Frohling S, Dunn IF, Schinzel AC, Barbie DA, Kim SY, Silver SJ, Tamayo P, Wadlow RC, Ramaswamy S et al (2009) Synthetic lethal interaction between oncogenic KRAS dependency and STK33 suppression in human cancer cells. Cell 137(5):821–834
99. Possemato R, Marks KM, Shaul YD, Pacold ME, Kim D, Birsoy K, Sethumadhavan S, Woo HK, Jang HG, Jha AK et al (2011) Functional genomics reveal that the serine synthesis pathway is essential in breast cancer. Nature 476:346–350
100. Meacham CE, Ho EE, Dubrovsky E, Gertler FB, Hemann MT (2009) In vivo RNAi screening identifies regulators of actin dynamics as key determinants of lymphoma progression. Nat Genet 41(10):1133–1137
101. Zender L, Xue W, Zuber J, Semighini CP, Krasnitz A, Ma B, Zender P, Kubicka S, Luk JM, Schirmacher P et al (2008) An oncogenomics-based in vivo RNAi screen identifies tumor suppressors in liver cancer. Cell 135(5):852–864
102. Dekanty A, Romero NM, Bertolin AP, Thomas MG, Leishman CC, Perez-Perri JI, Boccaccio GL, Wappner P (2010) Drosophila genome-wide RNAi screen identifies multiple regulators of HIF-dependent transcription in hypoxia. PLoS Genet 6(6):e1000994
103. Bergwitz C, Wee MJ, Sinha S, Huang J, DeRobertis C, Mensah LB, Cohen J, Friedman A, Kulkarni M, Hu Y et al (2013) Genetic determinants of phosphate response in Drosophila. PLoS One 8(3):e56753
104. Stotz SC, Clapham DE (2012) Anion-sensitive fluorophore identifies the Drosophila swell-activated chloride channel in a genome-wide RNA interference screen. PLoS One 7(10):e46865
105. Toret CP, D'Ambrosio MV, Vale RD, Simon MA, Nelson WJ (2014) A genome-wide screen identifies conserved protein hubs required for cadherin-mediated cell-cell adhesion. J Cell Biol 204(2):265–279
106. Cheung HW, Cowley GS, Weir BA, Boehm JS, Rusin S, Scott JA, East A, Ali LD, Lizotte PH, Wong TC et al (2011) Systematic investigation of genetic vulnerabilities across cancer

cell lines reveals lineage-specific dependencies in ovarian cancer. Proc Natl Acad Sci U S A 108(30):12372–12377

107. Birsoy K, Possemato R, Lorbeer FK, Bayraktar EC, Thiru P, Yucel B, Wang T, Chen WW, Clish CB, Sabatini DM (2014) Metabolic determinants of cancer cell sensitivity to glucose limitation and biguanides. Nature 508(7494):108–112
108. Goidts V, Bageritz J, Puccio L, Nakata S, Zapatka M, Barbus S, Toedt G, Campos B, Korshunov A, Momma S et al (2012) RNAi screening in glioma stem-like cells identifies PFKFB4 as a key molecule important for cancer cell survival. Oncogene 31(27):3235–3243
109. Colombi M, Molle KD, Benjamin D, Rattenbacher-Kiser K, Schaefer C, Betz C, Thiemeyer A, Regenass U, Hall MN, Moroni C (2011) Genome-wide shRNA screen reveals increased mitochondrial dependence upon mTORC2 addiction. Oncogene 30(13):1551–1565
110. McCleland ML, Adler AS, Deming L, Cosino E, Lee L, Blackwood EM, Solon M, Tao J, Li L, Shames D et al (2013) Lactate dehydrogenase B is required for the growth of KRAS-dependent lung adenocarcinomas. Clin Cancer Res 19(4):773–784
111. Gerlinger M, Santos CR, Spencer-Dene B, Martinez P, Endesfelder D, Burrell RA, Vetter M, Jiang M, Saunders RE, Kelly G et al (2012) Genome-wide RNA interference analysis of renal carcinoma survival regulators identifies MCT4 as a Warburg effect metabolic target. J Pathol 227(2):146–156
112. Pan J, Zhang J, Hill A, Lapan P, Berasi S, Bates B, Miller C, Haney S (2013) A kinome-wide siRNA screen identifies multiple roles for protein kinases in hypoxic stress adaptation, including roles for IRAK4 and GAK in protection against apoptosis in VHL-/- renal carcinoma cells, despite activation of the NF-kappaB pathway. J Biomol Screen 18(7):782–796
113. Miller JC, Holmes MC, Wang J, Guschin DY, Lee YL, Rupniewski I, Beausejour CM, Waite AJ, Wang NS, Kim KA et al (2007) An improved zinc-finger nuclease architecture for highly specific genome editing. Nat Biotechnol 25(7):778–785
114. Miller JC, Tan S, Qiao G, Barlow KA, Wang J, Xia DF, Meng X, Paschon DE, Leung E, Hinkley SJ et al (2011) A TALE nuclease architecture for efficient genome editing. Nat Biotechnol 29(2):143–148
115. Jinek M, Chylinski K, Fonfara I, Hauer M, Doudna JA, Charpentier E (2012) A programmable dual-RNA-guided DNA endonuclease in adaptive bacterial immunity. Science 337(6096):816–821
116. Mali P, Yang L, Esvelt KM, Aach J, Guell M, DiCarlo JE, Norville JE, Church GM (2013) RNA-guided human genome engineering via Cas9. Science 339(6121):823–826
117. Cong L, Ran FA, Cox D, Lin S, Barretto R, Habib N, Hsu PD, Wu X, Jiang W, Marraffini LA et al (2013) Multiplex genome engineering using CRISPR/Cas systems. Science 339(6121):819–823
118. Shalem O, Sanjana NE, Hartenian E, Shi X, Scott DA, Mikkelsen TS, Heckl D, Ebert BL, Root DE, Doench JG et al (2014) Genome-scale CRISPR-Cas9 knockout screening in human cells. Science 343(6166):84–87
119. Wang T, Wei JJ, Sabatini DM, Lander ES (2014) Genetic screens in human cells using the CRISPR-Cas9 system. Science 343(6166):80–84
120. Koike-Yusa H, Li Y, Tan EP, Velasco-Herrera Mdel C, Yusa K (2014) Genome-wide recessive genetic screening in mammalian cells with a lentiviral CRISPR-guide RNA library. Nat Biotechnol 32(3):267–273

Index

A
Acidity, 1, 21, 28
  as enviornmental cues and stimuli, 2
  titratable, 29
  tumor, 147
Acid-sensing, 1, 3
Acid-Sensing Ion Channels (ASICs), 3, 32, 46, 125
Arrestin domain containing 4 (ARRDC4), 70, 76, 78, 82, 84, 86, 141

B
Bis-2 [5-phenylacetamido-1,3,4-thiadiazol-2-yl] ethyl sulfide (BPTES), 102, 104, 105

C
Carbohydrate response element binding protein (ChREBP), 75, 77, 87
Carbonic anhydrases and voltage-gated proton channels, 123
Clustered, regularly interspaced, short palindromic repeats (CRISPR), 150

F
Functional genomic screens, 150

G
G2A (GPR132), 46, 55–57
Genomic, 116
  analyses, 151
  instability, 5
Glutamate, 18, 28, 71, 101–103, 107, 108, 146
Glutamate dehydrogenase (GDH), 87, 105, 109, 112, 144
Glutaminase (GA), 3, 75, 80, 87, 101, 103, 105, 109
Glutamine, 3, 71, 74, 75, 79–81, 91, 101, 109, 110, 145
  metabolism, 102, 147
  renal metabolism of, 105–107
G protein-coupled receptor 4 (GPR4), 3, 36, 47–50
G-protein coupled receptor (GPCR), 2, 3, 28, 36, 50
  signaling, 87, 88

H
Human antigen R (HuR), 113, 115
Hyperpolarization-activated Cyclic Nucleotide-Gated (HCNs) channels, 33, 35

L
Lactic acidosis (LA), 2, 4, 5, 70, 90, 136–140, 143, 147, 148
  metabolic symbiosis driven by, 74, 75

M
Max-like-protein X (MLx), 75, 76
  complexes, 78, 79, 81
Metabolic checkpoints, 3, 88
Metabolomics, 5, 116, 147, 151
MondoA, 70, 75–78, 81, 82, 91
Monocarbonate/bicarbonate transorters, 123

N
Nociception, 31
  acid, 31

O
OGR1 (GPR68), 46, 50–53

J-T. A. Chi (ed.), *Molecular Genetics of Dysregulated pH Homeostasis*,
DOI 10.1007/978-1-4939-1683-2

P
Phosphoenolpyruvate carboxykinase (PEPCK), 3, 109
Polycystic kidney disease (PKD)
PKD2L1/PKD1L3, 32–34, 38

R
RNA interference (RNAi), 5

S
Sour sensing, 32, 34, 38
many paths to, 38, 39
Synthetic lethality, 5, 148, 151

T
TDAG8 (GPR65), 46, 53–55
Thioredoxin interacting protein (TXNIP), 70, 78–80, 82, 83, 88, 89, 91, 143
other stresses that regulate, 80, 81
Transient receptor potential vanilloid 1 (TRPV1), 16, 21, 54, 125
Transient receptor potential vanilloid (TRPV), 125
Two-Pore Domain Potassium (K2P) channels, 33, 36, 39

V
Voltage gated chloride channels (ClC), 16
V-type ATPases, 123